# HALLEY HARPER,
# SCIENCE GIRL EXTRAORDINAIRE:
# THE FRIENDSHIP EXPERIMENT
## BOOK 2

Walla Walla County
Rural Library District

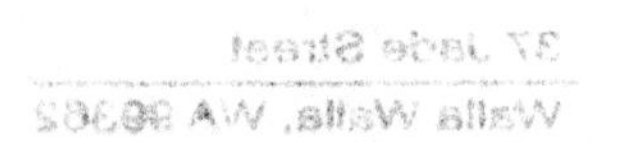
37 Jade Street
Walla Walla, WA 99362

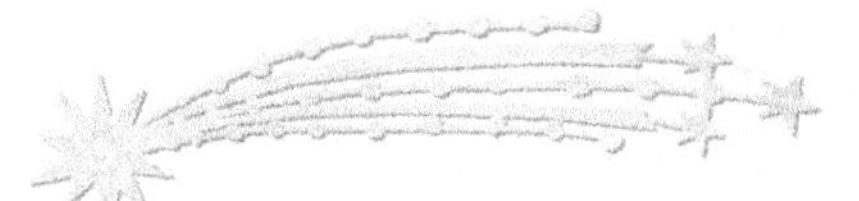

# HALLEY HARPER
## SCIENCE GIRL EXTRAORDINAIRE
# THE FRIENDSHIP EXPERIMENT

TRACY BORGMEYER

ILLUSTRATED BY
MELANIE CORDAN

www.TracyBorgmeyer.com

www.melaniecordan.com
Published by Tandem Services Press
www.Tandemservicesjvk.com



ISBN 978-1-7325285-1-2

Library of Congress Control Number: 2018908293

Suess, Dr. *Bartholomew and the Oobleck.* New York: Random House, 1949. Print.

*This one is for my cousins. Childhood memories are never very far away.*

*Science is like magic…but real.*

~ Author Unknown

# TABLE OF CONTENTS

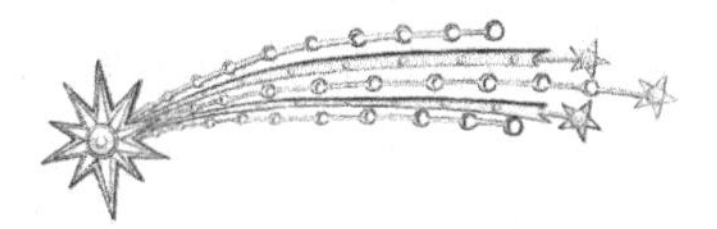

# CHAPTER 1
# A MIDNIGHT EXPERIMENT

SHHHHHHHH!

Halley Harper peeked at her best friend Gracelyn with her finger over her mouth as the two of them crept down the dark hallway. A middle-of-the-night science experiment was going to be so amazing. Halley's heart pounded thinking about how she had never done an experiment when everyone else was asleep. When she imagined how tasty it was going to be, her stomach let out a loud gurgle.

"Why does your stomach have to growl so loud?" Gracelyn whispered, peering around the corner towards the kitchen. "How come I get the feeling that I'm going to get in trouble for doing another one of your science experiments? We are going to get in so much trouble if we get caught here in the middle of the night. Mom will never let us have a sleepover again."

"Oh come on, Gracelyn!" Halley grinned, hugging the bag of marshmallows closely. "It will only take a few seconds in the microwave. And besides, s'mores are a perfect snack for a best friend sleepover."

"It does sound good, but let's hurry. I don't want to wake up my mom." Gracelyn tiptoed towards the kitchen while Halley followed.

The only light in the kitchen was the eerie green light from the microwave clock. It said 12:00. It was so quiet that the only sound was the big wall clock ticking.

"Ooh, it's midnight!" Halley whispered and followed Gracelyn to the microwave, bumping into her once before they got there. "We need a paper towel to put the marshmallows on. Quick, before we turn into pumpkins!" Halley knew Gracelyn would appreciate the princess reference.

"Wait, is this going to make a huge sticky mess?" Gracelyn stopped, tilted her head, and looked at her skeptically.

"Gracelyn." Halley grabbed her best friend by the arm and turned her to look at her eye to eye. "Have I ever gotten you in trouble?"

Gracelyn stared at her with one eyebrow raised.

"Okay, don't answer that question." Halley looked down. Gracelyn knew all the details about her science disasters. Well, they weren't disasters, just experiments that needed tweaking.

"Hurry up and let's just do this. You've made me hungry for s'mores!" Gracelyn looked over her shoulder and grabbed a paper towel to place in the microwave.

"Great. Okay, here is one marshmallow for you

and one for me." Halley opened the door and put the marshmallows in the microwave, quietly closed the door, and quickly dialed in 60 seconds. Before she pressed start, she turned to explain the science. "So what we will see is when the air trapped inside the marshmallows heats up, it will want to expand..."

"Yeah, yeah, hurry up and press start, Halley, and tell me the science when we get back to my room!"

Halley pressed the On button and the microwave oven came to life slowly whirring and turning the marshmallows. Nothing was happening yet.

Gracelyn peered in the microwave. Halley marveled at how Gracelyn's rainbow nightgown sparkled in the yellow light. She always looked so magical, but tonight she also seemed nervous standing in the kitchen at midnight. Halley, on the other hand, crossed her arms over her short-sleeve green **chemistry** T-shirt. She wasn't sure if she was shivering at how cold the house was or how excited she was thinking about eating warm gooey marshmallow s'mores.

The marshmallows were slowly getting bigger and bigger as they turned around and around in the microwave. They looked like giant sugar snowballs. "Whoa, look at the marshmallows expanding, Gracelyn!" Halley's eyes widened as she leaned closer to the microwave.

"Yeah, yeah, let's hurry up, Halley. When are they going to be done?" Gracelyn bit her fingernails and peered over her shoulder.

"Don't be so nervous." Halley touched Gracelyn's shoulder and turned back toward the marshmallows. "Only 10 more seconds." How big were they going to get?

"Oh no," Halley said under her breath. Could the marshmallows escape from the microwave? Halley didn't want Gracelyn to doubt her experiment.

"Oh no, what?" Gracelyn said. "Make it stop, Halley! Please don't make a mess."

"Just a few more seconds, Gracelyn." Halley breathed in the warm marshmallows. The smells triggered memories of summer camp that flooded over her and reminded her of how happy they were last summer at Camp Eureka.

"Take it out, take it out! I don't want it to catch on fire." Gracelyn pleaded as a charred hole started forming on one side of a marshmallow.

"It's almost done, just wait, 5,4,3,2,1." Halley counted down, mesmerized at the science happening right there in the microwave.

Suddenly, a bright fluorescent kitchen light flashed on. The girls turned around and gasped.

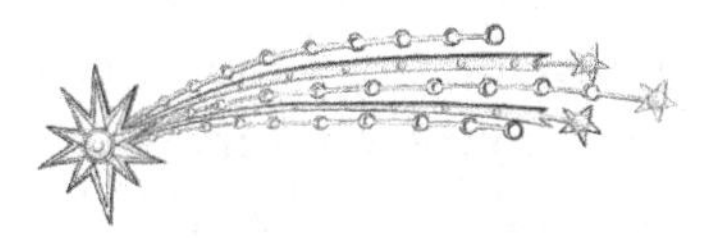

## CHAPTER 2
## CAUGHT GOOEY-HANDED

"WHY ARE YOU GIRLS UP SO late?" Ms. Dee squinted, looking back and forth between them for an answer. She pulled her bathrobe around her and tried to pat down her curly hair that was sticking straight up.

"Oh, hey, Mom." Gracelyn looked down and poked her toe into the kitchen tile grout. "We were just a little, uh, hungry."

"Ms. Dee," Halley gushed, still amazed at the results of her experiment. "You would absolutely love the science experiment we just did." She reached for the microwave door.

"We just made our own snack, Mom," Gracelyn interrupted. "Isn't that great?" Gracelyn gave Halley a wide-eyed don't-go-there stare and batted her hand down from the microwave door.

"Well, okay, girls, that's fine, but you need to run off to bed. It's late. By the way, are you planning on coming to our Halloween party, Halley? The garage will be decorated like a mad science lab and you could do one of your experiments to show the kids."

A mad science lab at Halloween? Halley clapped her hands together beaming inside. "Of course, Ms. Dee, that sounds amazing, and of course I'll help. What experiment do you want me to do?" Halley was giddy and more awake now than she should be in the middle of the night.

"Well, it's late, and we can talk about the details later." Ms. Dee yawned. "You girls run along and get into bed. You'll be like zombies tomorrow if you don't get some rest!" She gave Gracelyn an I-mean-business stare and turned, padding off.

"That was a close one." Gracelyn breathed a sigh of relief, "I can't believe she didn't ask what we were making!"

"Why are you so nervous, Gracelyn?" Halley questioned. "My mom is the one that flips out over messes, especially ones in the kitchen."

"It's hard to understand my mom sometimes." Gracelyn sighed. "She can be very protective of me. I guess because I'm her only child."

"Well, we are practically sisters, aren't we?" Halley admitted. Gracelyn may be an only child, but Halley only had a brother so her best friend was the closest thing to having a sister. The girls looked at each other and nodded in agreement. "Who cares about our moms?" Halley grinned and both girls looked toward the microwave at the same time.

Halley licked her lips. "Let's have some s'mores in our tent!" She opened the microwave, wrapped up the deflated gooey marshmallows in the paper towel, and squealed, "I've got the graham crackers and chocolate in my backpack!"

The girls ran across the kitchen skidding like sock-footed ice skaters on top of the freezing cold tiles.

Then they bounded down the carpeted hallway, muffling their giggles, and jumped into the blanket tent that took up almost all of Gracelyn's room.

"It smells just like Camp Eureka, doesn't it?" Halley breathed in the marshmallow's scent before she pulled it off the paper towel. The marshmallow looked like yummy goo when she sandwiched it between two graham crackers and a Hershey kiss. She passed the first assembled s'more to Gracelyn, because that's what best friends do, then sat back licking her fingers and peeked out the blanket tent to see the twinkling lights hung around the room left up since last Christmas. Halley squinted and imagined it was the Texas night sky at camp.

"Doesn't my room look like a perfect starry night in an enchanted forest?" Gracelyn blew a few graham cracker crumbs out of her mouth then giggled at how funny that was. "You are right, that experiment was worth it to have s'mores in the middle of the night."

"You're so funny, Gracelyn." Halley shoved the next s'more in her mouth and snuggled down in her glow-in-the-dark constellation sleeping bag. "I was thinking your room looks like Camp Eureka when we sat around the campfire after doing all the cool science challenges."

"Okay, so tell me now why the marshmallow blew up like that." Gracelyn rolled over and put her hands under her face waiting for Halley's answer.

Halley reached over and grabbed a fresh marshmallow and a flashlight. She shined the flashlight first under her chin. "A long time ago, a scientist in France found that a gas wants to expand when it gets hotter. Today scientists call it **Charles's Law**."

"Marshmallows aren't filled with gas, are they?" Gracelyn inquired nibbling on the corner of another graham cracker.

"Sure they are. Marshmallows are made of sugar and trapped air!" Halley said excitedly. "The microwave makes the trapped air hot, and the air expands causing the marshmallow to blow up!"

"Oh Halley, where do you get this stuff?" Gracelyn stared at Halley, shaking her head.

"Mostly I just ask a bunch of questions or read it in my favorite magazine, *Empower with Science*." She motioned over to the rolled-up magazine sticking out of her red backpack.

Gracelyn sat down behind Halley and started braiding her long brown hair. "Hey, speaking of *Empower* magazine, didn't you submit an article about summer science for the magazine's writing contest? Did you ever hear anything back from them?"

Halley thought back to the article she had worked on after Camp Eureka last summer. "No, I haven't heard from them. Oh well, no big deal if I don't win the contest." Halley lied. She always liked to win, and Gracelyn knew that about her.

"Well, I know you, Halley, and if you put your mind to it, one day you will be a published author because you are my best friend ever. Besides, who wouldn't want to publish an article making science as magical as you do?"

"Oh thanks, Gracelyn. You are always so sweet to me." Halley blushed.

"Actually, if you could just invent a machine that turns everything into marshmallows and candy then write about it, then that will always be the winner no matter what. Then we can eat as much sugar as we want without Mom finding out!"

"I'll have to write about that invention in my diary"—she reached for a pen and her diary in her backpack—"so I won't forget, then I'll make a ton of candy for you to eat."

Gracelyn finished braiding Halley's hair and adjusted half of the best friends necklace they were both wearing so the clasp would be in the back.

"Goodnight, Halley. The s'mores were delish." Gracelyn snuggled down in her sleeping bag, zipped it up, and covered her eyes with her rainbow sleeping mask that said Sweet Dreams.

Halley turned over, pointing her flashlight at her diary. She never wanted to forget how magical science could really be especially with her best friend.

*Dear Diary,*

*I love s'mores, and I love that Charles's Law helps me make my favorite part of the s'mores—the gooey marshmallow part. But why was Gracelyn so nervous about Ms. Dee seeing us making them in the middle of the night? Gracelyn was acting funny.*

*I can't wait for Halloween. I wonder what would*

*be the perfect Halloween experiment. Maybe I should ask Grammy. She'll know.*

*Love, Halley*

*P.S. Don't forget to invent a candy machine one day for Gracelyn.*

*P.P.S. Gracelyn is funny when she snores!*

# CHAPTER 3
# MONSTER TOOTHPASTE

HALLEY RACED UP THE SIDEWALK and stopped in front of the haunted mansion. The air smelled like a mixture of cinnamon sticks, candy corn, and dried autumn leaves. Halley adjusted her white lab coat, green safety goggles, and red backpack. She smoothed her tangled ponytail with one hand and clutched a giant beaker in the other.

She climbed the steps to the mansion and noticed cobwebs stretched over the bushes. A cloud passed over the big autumn moon. Black bats hanging from the trees whipped about in the cool breeze. She shivered and was glad her Halloween costume involved a coat.

Jack-o-lanterns glowed on the porch. Skeletons sat in chairs holding a sign that read, "Enter at your own risk." Eerie purple, orange, and green lights were dancing around, and she could hear ghostly noises inside.

Halley hesitated before going in. She didn't like haunted houses with things that jumped out and scared her. Taking a deep breath, she pressed the doorbell and braced herself.

Ding dong!

An evil laugh cackled all around her. The skeleton's teeth chattered, and a figure floated behind the glass front door.

"What's the secret password?" the figure cried out.

"The secret password? Um, trick or treat?" Halley gulped.

The doorknob slowly turned, and a unicorn horn poked out the door. A girl with a purple glittery face squealed and leapt out to hug Halley.

"Gracelyn!" Halley shouted, relieved. "I love how your unicorn costume turned out!"

Gracelyn beamed, adjusting the horn on her headband. "And, of course, you are a scientist!"

"At your service." Halley curtsied, holding her lab coat like a dress.

"Ooh! Is that for the experiment you're going to do?" Gracelyn touched the giant beaker. "Hurry, come on in! We've been waiting for you!" She turned around. There was a purple and pink unicorn tail pinned to the back of her pants. Her hair was braided in a French braid down her back ending in a big white bow.

"Your mom is so creative!" Halley admired Gracelyn's costume and her house that had been transformed into a haunted mansion.

"Hi, Halley!" Someone with a green face dressed all in black and wearing a giant pointy hat sang out in the kitchen. She was stirring a bubbling, giant metal bowl. "Would you like a bit of my witch's brew?"

"Nice touch, Ms. Dee! Is that dry ice?" Halley admired the purple potion and put her giant beaker on the kitchen table.

"I thought I'd try out a bit of science myself this year!" Ms. Dee grinned. "Thank you for agreeing to do some mad science for our haunted mansion!" She handed her a cup of the purple brew and adjusted the fake wart on her nose.

"Have you tried this experiment before?" Gracelyn said with her hands on her unicorn hips skeptically. "We don't need any disasters at the haunted mansion." She elbowed Halley, but she was smiling.

"I've tried it at my grammy's house before. She gave me everything I need to do Monster Toothpaste! I just need a bit of dry yeast. Do you have some?" Halley took her backpack off and peered inside, checking her ingredients.

"Of course! Gracelyn, show Halley where we set up the science lab. The kids will be coming in soon. I'll bring you the yeast." Ms. Dee hurried into the pantry.

Halley grabbed her giant beaker and ran after Gracelyn, who was skipping through the haunted mansion. They stepped out into the garage decorated like a science lab. Fake hands, eyeballs, and brains were floating in jars. A life-size Frankenstein was standing in the corner. Skeletons hanging on metal frames, were grinning like they were patiently waiting to watch the experiment. A glow-in-the-dark sign taped to a table said, "Halley Harper, Science Girl Extraordinaire."

"You're going to love this!" Halley giggled, unpacking her backpack on the station table. "It really made an oozy explosion at my grammy's house!"

"Did you say explosion?" Gracelyn questioned. "No one is losing an eyebrow tonight, are they?" She covered her eyebrows and stuck her tongue out, making a silly face.

"Not today! This is going to be super fun, you'll see!" Halley smiled and thought back to Camp Eureka. What was Ms. Mac, the zany, one-eyebrowed camp co-director, doing on Halloween? And it was her turn to write back her friend Nathan. She would ask him what his Halloween costume was this year.

"Here's the yeast!" Ms. Dee opened the garage door and tossed the yellow packets to the girls. "I'll check on you a bit later. I've got to let some guests in."

"What is the yeast for, anyways?" Gracelyn jumped up and down with excitement. "Are we going to make bread too?"

"Nope, Grammy said it's the **catalyst**!" Halley set up the supplies from her backpack.

"Did you say cat with a lisp?" Gracelyn grinned. "How is your cat, anyways?"

"**Atom** the Cat is awesome, but now is not the time to talk cats or catalysts. We don't have much time." Halley motioned to the kids piling into the mad science lab. "Here, safety first." Halley tossed Gracelyn some safety goggles.

Halley grinned. This was the first unicorn she'd ever seen with safety glasses on. Kids started watching what the girls were doing.

The girls stood behind the table while Halley poured her grammy's hydrogen peroxide into the beaker. She passed Gracelyn a bottle of dish soap. "Put a few squirts of soap into the beaker while I mix the yeast and water."

"By the way, what makes this Monster Toothpaste?" Gracelyn asked as she squirted the soap into the beaker.

"Well, when we add the yeast, it will look like someone is squeezing a giant monster-sized tube of toothpaste out. I'm adding some food coloring to make it fun." Halley squeezed the drops in, and a spooky green color dispersed like an alien hand swishing around in the hydrogen peroxide.

Kids who had formed around the station pounded their fist on the table while chanting, "Monster. Toothpaste. Monster. Toothpaste."

Halley felt the electricity in the air and wiped her sweaty hands on her lab coat. She hoped this worked the way it did at Grammy's house.

"What are you waiting for? The crowd is getting anxious." Gracelyn looked wide-eyed.

Halley picked up the yeast-and-water mixture. Her hand was shaking a bit. Then she quickly dumped it in the hydrogen peroxide.

The green liquid started foaming.

Halley blinked.

Suddenly, green foam shot violently out the beaker. A few kids gasped and jumped away from the table. Gracelyn started screaming as Monster Toothpaste blasted up into the air hitting the ceiling before raining down.

The green foam landed on Halley's head. It oozed down her hair and onto her forehead. She wiped off her safety goggles and looked around. The Monster

Toothpaste, splattered all over the mad science lab, started smoking.

# CHAPTER 4
## HALLEY'S NEW 'DO

THE SMELL OF CHEMICAL FILLED Halley's nostrils, reminding her of a super clean doctor's office right before a visit. Her head started to feel crawly then it itched. A lot. She reached up to scratch her scalp, and a small clump of hair came out in her fingers.

"Oh no, that's not good." Halley picked the strands of hair out of her fingernails. *I've really done it this time.*

Most of the kids watching the experiment had screamed and left the garage, leaving Halley in watery green puddles splattered all over the floors and globs of green oozing down the walls. Gracelyn had run to get help.

"What happened? Are you alright?" Ms. Dee shouted, returning with Gracelyn. She took off her witch's hat and surveyed the garage, then looked at Halley. "What have you done, Halley?"

"I was just doing the Monster Toothpaste experiment." Halley reached up to rub her prickling scalp. Gracelyn appeared to have dodged the downpour of green ooze.

"What is this stuff you brought?" Ms. Dee reached down to inspect the brown-labeled bottle. "Why on earth would your mom let you get into this chemical?"

"My mom doesn't know that my grammy gave it to me." Halley was only half listening. Her scalp was on fire with prickly chemical bugs.

"We need to get you home so you can wash your hair right away. I wonder if any other kids were sprayed with this stuff." Ms. Dee scanned the garage.

"I'll walk her home!" Gracelyn offered reaching to put her arms around Halley's shoulders.

"No, I think it's best you stay here with your dad, Gracelyn." From the tone of her voice, Ms. Dee sounded like she was trying to protect Gracelyn from a dangerous zoo animal that had escaped its cage. Except that dangerous animal was Halley.

Ms. Dee tossed the hydrogen peroxide and beaker into Halley's red backpack and whipped it over her shoulder. She looked over at Gracelyn whose eyes were welling up with tears, and Halley felt sick to her stomach.

"Is Gracelyn going to be okay?" Halley mumbled as she and Ms. Dee left the garage and walked down the street to her house. The brisk Halloween night made Halley's hair feel a little bit better, but it was feeling drier and more brittle by the moment. It seemed like any minute it was going to just crack right off. She needed to wash it out of her hair quickly when she got home. Her mind wanted to figure out what happened, but her hair preoccupied her thoughts with itching. Did she grab the wrong bottle at her grandma's house?

"Thank goodness you both were wearing safety glasses." Ms. Dee fumed. "This was alarming, Halley, and I'm just glad no one got hurt in your little experiment. I should have never asked you to come over and do this. Maybe your mom should have come and supervised you."

Halley's eyes bulged and stung at the same time. She was trying to process what went wrong in the experiment. All she was doing was trying to help. Ms. Dee was making her feel like she did this on purpose.

"I'm so sorry, Ms. Dee." Halley attempted an apology but was too distracted. "My hair just doesn't feel right." Were her ears starting to itch too? If she didn't look like a crazy mad scientist before, she did now since her hair was frizzing up and cracking off.

When they finally got to Halley's house, the porch light was off but Ms. Dee marched up and knocked, sounding like a late-night, impatient trick-or-treater.

The door opened. Mom already had her comfy clothes and slippers on and had thrown her long hair into a bun.

"Isn't it a little late for trick or treating?" Mom snapped then saw who it was. "Oh, hi Dee. Is everything alright?"

"Halley needs to have her experiment washed out of her hair, and I need to get back to clean up the mess that was made." Ms. Dee's voice was abrupt as she handed Mom the red backpack.

"Oh my gosh! Halley, what happened to your

hair?" Mom touched Halley's hair, holding it like the cobwebs that were draped over the front shrubs.

"Halley's little science accident at our house tonight caused chemical foam to rain down in our garage during the party. I really need to go and make sure no one else gets into it." Ms. Dee breathed heavily. "And listen, I think *you* need to pay a little more attention to what Halley's got in her experiment inventory."

"Well I'm sure there is a logical explanation for this." Was Mom actually defending her?

"Please, Mom, let's just get this stuff out of my hair." Halley didn't feel like being polite anymore with her scalp on fire.

"And you know what?" Ms. Dee raised her chin. "I think Halley and Gracelyn need a little bit of space. There have been too many of these near-miss accidents, in my opinion."

"What? I can't see Gracelyn anymore?"

"Well, I'm sorry she made such a mess," Mom replied. "Do you want me to come over and help you clean up?"

"No, I'll take it from here." Ms. Dee quickly turned on her heel and stormed back to her house. Her witch's costume flying behind her.

Mom looked down and inspected Halley's hair. "What experiment did you do, and where did you get the ingredients to make such an explosion?" Mom closed the door and led Halley to the kitchen sink, not waiting for an answer.

Halley wriggled out of her lab coat, her mind racing in sensory overload.

Dad rounded the corner wearing one of his favorite Texas A&M shirts, flannel pants, and socks. "Halley, is that you? What happened to your costume and your hair?"

"Help me wash out her hair," Mom instructed, leading Halley to the kitchen sink and leaned her head over it while Dad rinsed out her ponytail and scalp with the sink sprayer. Halley saw strands of hair coming out in the drain, but at least she felt some relief. Mom came over with a towel, patting her hair dry, and led her to the kitchen barstool.

Halley sat and started sobbing. "I just wanted to have a cool experiment for the haunted house, and the kids were pressuring me, and I think I grabbed the wrong chemical in Grammy's kitchen. And my hair!" She reached up to her wet head. "What's going to happen to my hair?"

"We're just going to have to cut your hair and salvage what we can. I'll make an appointment for you to get it cut and styled." Mom patted her shoulder. Halley was so glad she wasn't getting in trouble for this with her mom.

Mom started combing through Halley's hair, and Dad reached down to inspect the brown bottle. "Concentrated hydrogen peroxide," he read. "Grandma must use this stuff in small amounts to dye her hair. You're lucky it didn't get on your eyebrows." Dad reached down and gently inspected Halley's face.

Halley remembered Ms. Mac, Camp Eureka's zany director, and her one eyebrow. It must have been a science battle scar.

Mom held up Halley's brittle ponytail, turning it

over in her hand before Halley felt the scissors clamp down on her hair, making an awful crunching sound. "You know," Mom said, "perhaps you need to focus less on science experiments and more on a less-disastrous hobby, like your piano lessons." She just kept cutting her hair shorter and shorter and more strands of hair fell around her.

Halley looked over at the piano and the little porcelain haunted house decoration that blinked when it lit up. It cast a warm glow over the piano. She thought about Gracelyn.

She started crying again remembering Gracelyn's face and how worried she looked.

"Don't worry." Dad squeezed her knee. "You'll be the talk of the town with your new hairstyle."

Her head felt lighter with less hair, but it still swirled with everything that just happened. She sat perfectly still while Mom swept hair off the neat and tidy kitchen floor. The light from the porcelain haunted house decorating the piano just blinked on and off, on and off. Surely Ms. Dee was just mad. How could she possibly keep Halley from seeing her best friend?

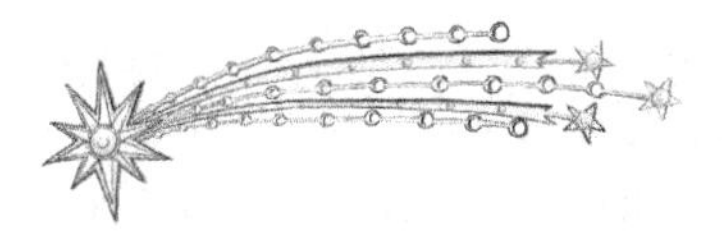

# CHAPTER 5
# HALLEY'S GOT MAIL

HALLEY GROGGILY OPENED HER EYES, forgetting for a moment what day it was. She rubbed them and rolled over in bed, looking out the window. It was still dark outside, and she could hear the wind blowing through the trees and the morning birds chirping. She stretched out, tucking her feet under the warm blankets and pulling the covers back up to her chin and closed her eyes.

Opening them quickly, she looked over in her room and saw the solar system she begged Mom to let her hang from the ceiling with string.

Was it all just a dream? She swallowed and felt her teeth with her tongue. She must have been so tired she forgot to brush her teeth the night before.

She sat up, stretched, and scratched her light-feeling head.

What happened to her hair? It was gone.

She jumped up, ran into the bathroom, and switched on the light. Squinting, she looked in the mirror. She put a hand to her hair again. There, looking back at her, was a long, lanky girl with puffy eyes and short hair rubbing her head.

It wasn't that bad. Her bangs laid across her forehead, and she tried to sweep them back behind her ear with no luck. But every time she looked in the mirror she would think of how Gracelyn looked on Halloween night standing in the green ooze. She had to find a way to talk to her again and apologize.

Halley took a deep breath and leaned closer to the mirror. Her hair would be a lot easier to take care of, and it wouldn't take forever to dry. She convinced herself that she would get used to her new easy-to-fix look. She grabbed her blanket, slung it around her shoulders like a cape, and padded down the carpeted hallway. It was the day after Halloween, and she had to get ready to go to school.

The kitchen light was on. Her five-year-old brother, Ben, still wore part of his Halloween costume, a magician's top hat and cape, with pajama bottoms and was inspecting his Halloween candy haul. Dad was drinking coffee and reading the newspaper.

"Hey there, Miss Disaster." Dad reached out and squeezed her arm. "How are you feeling this morning?"

"What happened to your hair?" Ben squealed before she could sit down at the table. "Was that part of your Halloween costume?"

"Ben," Dad warned, "leave Halley alone. She's been through a lot last night."

"Were you trying to be a boy for Halloween?" Ben

teased. "Ouch! Dad, don't kick me under the table!"

"I decided to go for a new look." She snatched a piece of candy from Ben's pile as retribution for his comment. "It's all the rage to have super short hair."

"Well, I think it looks great on you." Dad folded down the newspaper and smiled warmly. "It makes you look a lot older and more mature. How about some pancakes?" Dad threw a glance at Ben to stop the teasing. "Oh, and I've got something for you." He strolled over to the computer and reached for an envelope. "You've got mail!" He tossed an official-looking letter at her.

She loved getting mail. Who could possibly write her? She turned it over several times, inspecting the envelope. It had a logo of a beaker with flowers coming out of it and below it read, *Empower with Science* magazine.

"Oh my gosh," she whispered, starstruck. "It's from *Empower with Science* magazine." Halley thought about her favorite magazine and all the amazing ideas it gave her. Although sometimes it could get her in trouble, but for now all she could think about was the essay she mailed to them about her summer adventures at Camp Eureka. Dad was so proud of that essay. What if they were writing to say the article was going to be published!?

She turned over the envelope and slid her finger under the flap, careful not to rip the letter inside. Was this what opening a golden ticket for the chocolate factory felt like?

"What does it say?" Dad beamed, breaking her thoughts.

Ben was too busy stuffing more candy corn in his

mouth for breakfast to pay attention.

She pulled the letter out and carefully unfolded the typed paper.

> Dear Miss Harper:
>
> We regret to inform you that the "My-Summer Set in Motion" article you submitted to our magazine will not be published in *Empower with Science* magazine.

She looked up at Dad then blinked, trying to hold back the stinging tears. She reached up instead to twirl her hair but remembered most of it was gone.

That was just her luck. First her best friend, then her hair, now this? Couldn't the world give a ten-year-old a break?

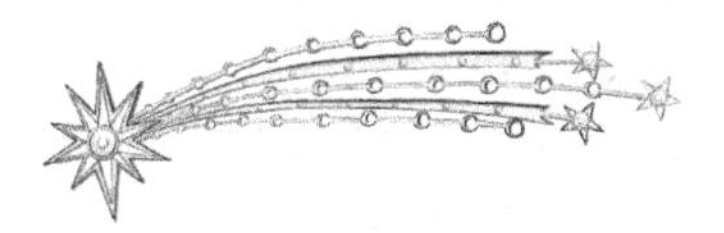

## CHAPTER 6
## KID SCIENCE REPORTER

HALLEY TOSSED THE LETTER ON the worn breakfast table and glanced back and forth between Dad and Ben. "Oh well," she muttered, "they aren't going to publish my article." She plopped down at the table, staring out the kitchen window, and tried to swallow back the tears forming.

Outside a mourning dove landed on her treelab playhouse and started singing. Halley dreamed of turning her treehouse into a time machine and going back before Halloween to fix all of this mess. But first she would write a letter to Gracelyn apologizing.

Suddenly, a little hand rumpled her hair, interrupting her inventive thoughts.

"Stop it, Ben! Don't touch my hair." She whined loud enough for Dad to hear.

"Ben, stop it!" Dad had a clenched jaw and a don't-mess-with-your-sister look.

"Oh, come on. I just wanted to know what it felt like! Her hair is all gone!" Ben scooted his chair toward the table, making a loud screeching sound and grabbed the maple syrup bottle turning it over and squeezing as hard as he could over a pile of candy corn.

"Gross, Ben. That is way too much sugar for breakfast." Halley crossed her arms.

Ben's sugary breakfast looked like a candy volcano erupting with syrup, threatening to engulf the tossed-aside letter.

Dad reached across the table, snatched the letter from the impending syrup lava flow, and scanned it.

A smile stretched across his face. "I think you should look at this letter again." He flattened the paper out and handed it to Halley.

She picked it up again, crossed her legs in the chair, and silently read.

> Dear Miss Harper:
>
> . . .
>
> However, we want to congratulate you and welcome you to our team as a kid science reporter. We have chosen you and one other lucky reader to investigate science topics for us.
>
> Statistics show that girls begin to lose interest in science at an early age and will avoid a science career simply because of the limitations that exist in her own mind. Solu-

tions for today's problems will come from tomorrow's science – quite possibly you and your friend's science. We at *Empower with Science* magazine believe that kids like you are part of this solution.

Your first assignment is to go back to Camp Eureka and interview the director, Ms. Mac, about why your friends should go to Camp Eureka this summer. We hope this will encourage girls to attend this wonderful summer camp and continue their interest in science.

Congratulations and good luck!

Sincerely,

Kelly Winchell
Editorial Director of *Empower with Science* magazine

Halley looked up from the letter grinning. "They picked me to be a kid science reporter and best of all—I'm going back to Camp Eureka to see Ms. Mac!"

"I think your luck is about to change!" Dad chuckled, reached over to hug Halley, and slid a big stack of pancakes in front of her. "Now come on, my little comet, and eat your breakfast. We've got to get you to school on time. After your piano lesson this afternoon, Mom is going to take you and Ben to get ready for his birthday party."

"Happy birthday to me...CHA CHA CHA!" Ben

sang with syrupy wet bits of candy corn and pancake in his mouth.

Halley snorted and rolled her eyes. She wasn't going to let Ben ruin this moment for her. She couldn't believe it. She was going back to Camp Eureka to see Ms. Mac again. She stuffed her mouth full of sugary warm pancakes. All of this news made her hair drama seem like a distant memory. Gracelyn would be so excited for her being selected as a kid science reporter. But who was the other reporter? Halley's heart raced with excitement.

Suddenly, the kitchen garbage disposal whirred on. Dad was cleaning up the breakfast dishes, but Halley had a flashback of her Oobleck experiment she poured down the drain last year.

"Want apples in your lunch?" Dad questioned.

"Um, sure, Dad." Her excitement changed quickly to worry. "Don't forget to squirt a bit of lemon juice on the apples. Lemon juice is **acidic** and it will keep the apples from turning brown!" Halley sang out, but then she remembered what Ms. Spark looked like getting stuck in Oobleck at camp.

Would Ms. Spark be there when Halley visited Camp Eureka for the interview? Was Ms. Spark going to seek her revenge because Halley revealed her plot last summer to close Camp Eureka forever?

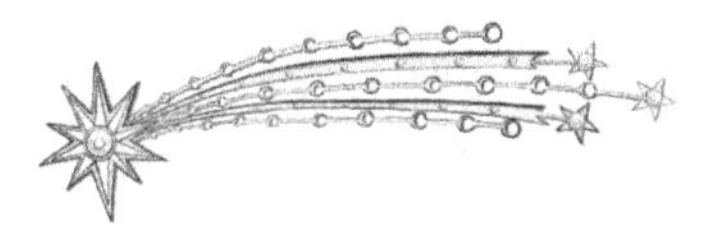

# CHAPTER 7
## MAGIC BIRTHDAY PARTY

HALLEY MANAGED TO MAKE IT through school and piano practice that day by pushing thoughts of Ms. Spark out of her mind. She was too busy thinking of a way to talk to Gracelyn, who hadn't shown up at school that day, and trying to avoid kids who were asking about her hair.

Now the bell on the door of the party supply store chimed loudly when her, Ben, and Mom stepped inside. Halley smelled balloons and leftover Halloween candy for sale. It was bad enough that she was being dragged along to the store, but now she had to be happy to celebrate her annoying little brother's birthday.

Ben reminded Mom he wanted a magic show party and ran ahead to the birthday party aisle. Mom clapped her hands with excitement and began to peruse the store. "Maybe we could hire a magician to perform at Ben's party." Mom said to herself as if making a mental note. She never got this excited when Halley asked for a science-themed birthday party.

Halley walked down the aisles behind Mom, brushing her hands over all the favor trinkets; slime containers, bouncing balls, and bubble gum. It vaguely reminded her of the General Store at Camp Eureka with its tasty science food and mind-boggling toys. The camp store seemed magical. This store seemed fake, cheap, and packaged.

She found some plastic bunny key chains and tossed it at Ben. "Hey Ben, you could pull this plastic rabbit out of a hat." Halley teased.

He stuck his tongue out at her and continued pulling plastic flowers out of a magician's wand.

"Welcome to Party Place," a saleslady sang out, sashaying down their aisle.

"Um, yes." Mom looked around, rubbing her hands together. "We are looking for ideas for a magic show birthday party. You know, paper plates, banners, and party favors."

"Oh, that sounds amazing. Now let me see..." the employee chimed. "I see you've found the right aisle. We have a ton of magic themed plates and paper goods." She swiveled on her heel and bent down to Halley and Ben's level and put her hands on her hips. "Now which one of your adorable little boys is this birthday party for?"

Ben snorted and peered over at Halley. Mom pursed her lips together and picked at her fresh pale pink fingernail polish waiting for Halley's response.

"I am NOT a little boy!" Halley scowled at the lady,

sternly crossing her arms in front of her. She felt storm clouds form in her eyes making them hot with tears.

"Halley, she thought you were a boy!" Ben busted out laughing loud enough for the whole store to hear.

"Oh, I am terribly sorry." The saleslady stammered, backtracking. "It was just that they look so much alike."

"That was a terrible recovery." Halley said narrowing her eyes.

"Oh no! This is for my son Ben's birthday party." Mom stopped behind Halley, placing her hands on her shoulders. "This is his sister, Halley."

Halley swiftly turned, breaking free of Mom's grip and started inspecting the mad scientist wigs hanging on the wall. She would never be the girly girl that her mom had always wanted.

"She can be a little sensitive," Mom lowered her voice, "and kind of a nerd too."

She couldn't believe her mom just told a complete stranger that she was a nerd. Mom wouldn't dare say that to her when she was a published kid science reporter.

On the car ride home Halley wondered what ingredients she could mix together to make hair grow faster. Just you wait, Halley seethed. She would never be mistaken for a boy again.

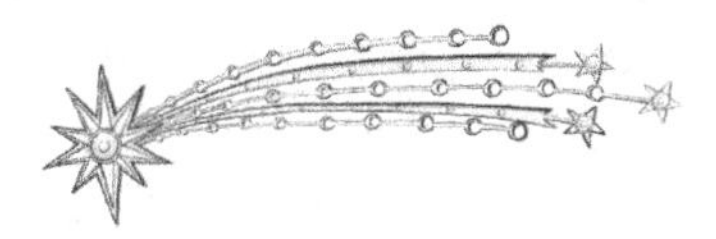

# CHAPTER 8
# GRAMMY

MOM'S PHONE RANG OVER THE car speakers as they bumped into the driveway. "Oh good, it's Dee." Mom muttered as she pulled into the garage and turned off the ignition.

"Well hi, Dee, thank you for returning my call. We wanted to check on Gracelyn. Halley said she wasn't at school today."

"She is fine, thank you for asking." Ms. Dee's voice boomed over the car speaker. "But seriously, don't you think Halley's experiments are starting to get out of hand? This one could have turned out very badly." Ms. Dee fumed.

"Well, Halley practically lost all of her hair in the accident," Mom explained.

"Perhaps you need to pay more attention to what she is getting into." Ms. Dee snapped.

Halley opened her mouth to try and protest.

"Well, perhaps you need to mind your own business!" Mom barked.

"I think it's best that they take a little break from each other and find some new friends."

Halley swallowed and sat speechless.

"Well, I never!" Mom looked down at her phone with stormy eyes and quickly hung up the phone.

Halley couldn't take it anymore. First, being called a boy; then hearing Mom and Ms. Dee fighting. She just wanted to be left alone, so she raced toward her treelab in her backyard. Maybe she could start on her super hair growth formula. She reached the top of the tree and opened the door to the treelab and slammed it behind her. Stepping inside, she plopped down on a bean bag chair and closed her eyes.

Halley thought back to the sleepover at Gracelyn's and how magical her room looked, back when they were still best friends. She felt something on her leg and jumped. It was just Atom rubbing up against her leg wanting to be petted. "I'm so glad you're still my friend, Atom." She loved her cat. How did he know when she needed a friend the most? Would Atom let her test her hair growth formula on him? But he was such a nice-looking, short-haired Siamese cat.

Knock, knock, knock.

"Go away, Ben, I'm busy," Halley snapped.

KNOCK, KNOCK, KNOCK! The knocking grew louder and more impatient.

Halley flung the door, "I said, go away, Ben! You are so annoying!"

"Well, hello, pumpkin! I think you and I have a little talking to do."

"Grammy?! What are you doing here?" Halley's emotions melted away as she looked into her grammy's soft warm eyes and understanding smile. Grammy always made everything feel better.

"Come into my treelab!"

"I don't mind if I do." Grammy ducked and entered into the playhouse that Halley had transformed into a sciene and craft lab. She surveyed the little room. "It's a bit cluttered, but you know that some of the most brilliant minds had messy rooms. Me? I'm not one for organization."

"I don't like to put things away because I want to see where everything is." Halley shrugged admitting to her messiness.

"Let me take a look at you. You've grown a foot since I've seen you last! And I adore your new hairdo." Grammy held Halley at arm's length to look at her.

"Did you hear how I got this short haircut?" Halley pushed her hair from her face and almost forgot that Grammy was partly to blame for the accident. "And a saleslady just thought I was a boy! I just want my long, pretty hair again."

"Your Dad told me all about it, which is why I'm here to say I'm sorry. I must have grabbed the wrong bottle of hydrogen peroxide to give you. I was going to use it to make myself a blonde, but turns out I like my hair the way it is. Will you ever forgive me?" Grammy stood with her chin tucked waiting for Halley to react.

"Of course, Grammy. I love you!" Halley melted into another big hug.

"Listen, the great thing about hair is that it will grow back. In the meantime, you can use this."

Grammy reached in her pocket and handed something to Halley.

Halley turned it over, lightly inspecting a hairpin with a small, sparkly comet sitting on the end. "Can you help me put it on?"

"Of course I will." Grammy smiled, cradling Halley's face then slipped the hairpin on her side-swept bangs. "This pin is to remind you that you are named after a shooting star, Halley Edison Harper, and you are a beautiful girl because prettiness on the inside always shines to the outside." Grammy wiped a tear that had formed at the corner of her eye and cleared her throat.

"Now, stand back and let me look at you. I've got to get running, but I wanted you to watch tonight for a meteor shower. These shooting stars are special because they are made up of dust from Halley's comet."

"I love you, Grammy." Halley fell into Grammy's arms and everything felt right. Her visit reminded Halley that she was named after a comet, and that no one could make her feel bad unless she let them—not Ben, not the saleslady calling her a boy, not Ms. Dee, and especially not her mom.

That night, long after Halley should have been asleep, she stared out her window watching the meteor shower. Everything seemed right again in Halley's world until she spotted a bright, slow-moving meteor lighting up the night sky. Suddenly, it broke off into two pieces that went flying in opposite directions before fizzling out in the night.

Halley looked over to her nightstand where she had a picture of her and Gracelyn at her last birthday party. Was that meteor that broke apart a sign that she and Gracelyn were meant to be apart forever? She was going to have to figure out a way to get Gracelyn to come to Camp Eureka that summer so they could be together again—no matter what.

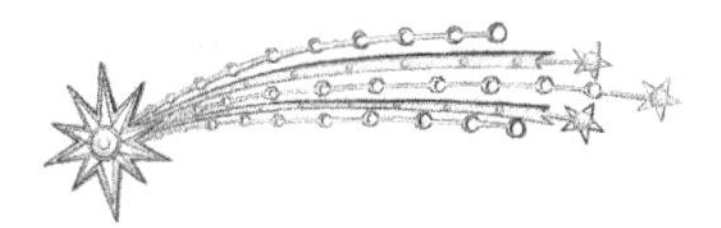

## CHAPTER 9
## WINTER AT CAMP EUREKA

Clouds of white exhaust drifted up from Dad's car into the cold, gray winter sky when they pulled onto the dirt road leading to Camp Eureka. The forest was bare without all the trees bursting with green and eerily quiet without wacky teen counselors wearing neon-green camp shirts to greet them. It was hard to imagine the warm summer memories when winter had taken over camp.

On the car ride there that morning from their home in Houston, Halley spotted an Alaska license plate and secretly wished she was playing the license plate game with Gracelyn. She did not want Dad to know how lonesome she felt going back to camp without her best friend, even if it was just for the afternoon.

At the end of the road, they rounded a corner and the main camp building suddenly came into view with a sign that read Energy Hall. Standing beneath the sign was a tiny lady—with wild, frizzy hair—smiling and waving to greet them. She wore a heavy army-green coat, scarf, and a woolen cap over her head. It was Ms. Mac.

Halley smiled and waved back.

"You gonna be okay interviewing Ms. Mac by yourself, Halley?" Dad put the car in park and leaned over, smiling reassuringly. His jacket made a warm crunching sound on the seat. "I'm so proud of my real-live kid science reporter. I'll see you in a couple of hours, Miss Disaster."

"Sure, Dad." Halley smiled, looking Dad straight in the eye and patting his shoulder. She didn't want Dad to know how nervous she was. "Don't worry, I know this place like the back of my hand."

Dad lowered his window as Ms. Mac walked up to the car and leaned in. "Hi, Halley! Great to see you again!" It was so cold that Ms. Mac's breath puffed like a cloud. "You ready to interview ol' Ms. Mac?" She raised her one eyebrow, and her eyes sparkled.

"Hi, uh, yes, I think I'm ready!" Halley clutched her diary close to her chest that she had written her first-time reporter questions in. She hadn't even let her Dad see what questions she was going to ask Ms. Mac.

"Halley has been talking nonstop about coming back to Camp Eureka. Thank you so much for meeting with her to help her on her assignment," Dad said warmly.

"Anytime," Ms Mac said. "This is going to be a piece of cake. Right, Halley? Why don't you go into the Hall through the front, and I'll meet you back in my office? I've got something to check on out here."

"Bye, Dad!" Halley grinned, pulling her red knitted hat down over her ears. Too bad she didn't have her long hair anymore to keep her neck warm. A wind tunnel formed after she opened the car door and whipped her short hair around. Halley blew her warm breath, and it made a little cloud of her own. This was cold even by Texas weather standards. Even her eyeballs felt cold. She blinked and the Energy Hall sign came into focus.

She turned and gave Dad a wave goodbye as he slowly drove away before she climbed the steps. Energy Hall had a soft, yellow glow coming from inside. Halley smiled. She remembered how much she loved that all of the camp landmarks were named after something in science: Energy Hall, Lake Archimedes, and Pie Are Square. It would be neat to have a Halley's Comet building one of these days.

She pulled open the heavy, metal door of Energy Hall and the first thing she noticed was the sweet smell of hot chocolate. A small bit of her nervousness melted away as she looked around the building and took in the familiarities of camp. She pulled off her red hat and rubbed her frigid hands together, noticing a light blinking on and off, on and off in Ms. Mac's office down the long hallway.

Halley lingered before walking down the hallway, remembering the first time she was here with Gracelyn. Her heart started aching again. It was almost Christmas, and she had not talked to her best friend in nearly two months! Had Gracelyn even enjoyed her Thanksgiving turkey?

The hallway was covered with pictures tacked on bulletin boards of kids at camp. She could spend hours looking over the last ten summers of Camp Eureka. She moved down the hallway inspecting each picture one slow step at a time. She was so engrossed in the camper's faces that she glanced away only when she stubbed her toe on the giant Newton's Cradle she and Gracelyn found last summer. She smiled and pushed the suspended balls, causing them to knock together gently.

She refocused on the sea of pictures and found last summer's camp: Set in Motion. There were pictures of the Newton's Laws challenges, balloon rockets, Unbalanced Rock, and the Maze of Science. Then she came to a picture of her team, Team Comet. She admired the image of her, Gracelyn, and Nathan running across the Oobleck pool they made for their end of camp Team Comet project.

How many pictures were there of Nathan, since he attended camp a few years before Halley? Halley wrote him a letter before coming to interview Ms. Mac. "Make sure you ask about the caves that the older kids get to explore," Nathan had advised her.

Then her eyes stopped on a fuzzy picture of a skinny lady with a pinched expression holding a clipboard tightly. It was Ms. Spark. Halley's stomach sank. What if she was here in this building with her right now? Would Ms. Spark say anything to her about her confession last summer of trying to end Camp Eureka?

Something looked out of place. Tucked behind the picture of Ms. Spark was an old, folded brochure that appeared to have been put there quickly. Halley lightly took it down from the bulletin board and opened up the brochure which read, "Camp Sokatoa: a Camp for Future Astronauts." What was this, and why was it here in Camp Eureka's pictures? Halley took it and slipped it in her diary, promising herself to ask Ms. Mac about it when she got the chance.

She had reached the end of the hallway, and Ms. Mac's office light was still blinking on and off behind a closed door. Halley knocked softly and could see a shadow passing across the frosted glass. Ms. Mac opened the door, and Halley was surprised that she was almost as tall as Ms. Mac was, with just a little less than a head to go.

"Halley Harper, well aren't you a sight for sore eyes!" And Ms. Mac pulled Halley in for a big bear hug. She was strong for such a tiny lady.

"Now, stand back, let me take a look at you." Ms. Mac leaned Halley back and inspected her. "I just love your new short hairdo!"

"Thanks. It was sort of a science accident."

"Well, I'm all too familiar with little science accidents." Ms. Mac leaned in close to Halley, raising her one good eyebrow. "Come on in. Take a load off!"

Halley stepped inside the office and could see that a small Christmas tree about her height was strung with white lights blinking on and off. As she got closer she noticed it was covered with other lights that bubbled, and beautiful white crystal snowflakes hung on the branches.

"What do you think of my tree? Christmas is always my favorite time of year." Ms. Mac sat at her desk and stared at the tree.

"Those are beautiful decorations." Halley wanted to touch the crystal snowflakes to see how they felt. Were those real crystals, and why did those lights bubble?

"Want to take a closer look?"

Halley went up and touched the bubbling lights first. They were warm. Then she poked at the tiny, perfectly white, sparkling crystals on the snowflakes. For a moment, she forgot why she was there.

"So, I hear you are a reporter now." Ms. Mac snapped her back to reality. "You are a gal with many talents, science and writing! Why don't you sit right down and ask me your questions?"

"Oh, right." Halley stepped over to the big comfy couch and plopped down. "I'm supposed to write about why my friends should come to Camp Eureka. So I came up with a few questions to ask you." Halley opened up her diary, and the old camp brochure fell on the floor. She looked up to see if Ms. Mac would be mad that Halley took it off the wall and put it in her diary.

Ms. Mac leaned forward and stood in one quick motion to inspect what fell on the floor.

"I, um," Halley stuttered, "was hoping you could tell me about this old camp brochure after our interview." She held up the map and took a deep breath. "What is Camp Sokatoa?"

Ms. Mac crossed her arms, staring at the brochure then at Halley. "I think this interview is over."

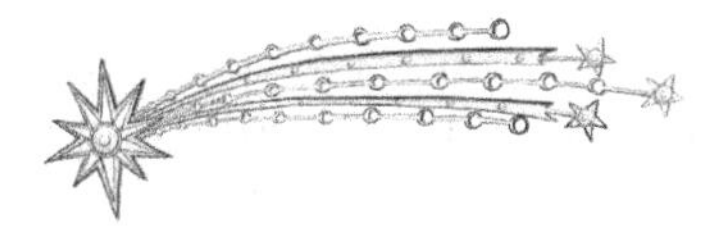

## CHAPTER 10
## A NATURE WALK

"WHAT? WHY?" HALLEY STAMMERED IN disbelief looking at Ms. Mac then back at her diary full of questions she had carefully written out. How would she ever complete her assignment if she didn't interview Ms. Mac?

"Why tell you about it when I can show you!" Ms. Mac leaned in closely, and a grin slowly turned up the corners of her mouth. Ms. Mac jumped up and grabbed her heavy coat, surprising Halley at the change of her mood. She wasn't upset at all; in fact she seemed excited. She quickly wrapped a scarf with little printed **molecules** all over it around her neck. "Well don't just sit there, we've got a lot to see! Grab that old brochure and help me with my camp inspection. There are only six short months until camp is in session!"

Halley sat there stunned. "Isn't it pretty cold out

there?" Halley's toes were still numb from the short walk from Dad's car to Energy Hall.

"Don't dilly dally, Halley!" Ms. Mac grabbed a clipboard and bolted out the door.

Halley had to run to catch up as they went out the back of Energy Hall and into the forest of Camp Eureka.

She shivered and looked around the camp grounds and took in the beautiful scenery. There in the middle of camp was Lake Archimedes. The wind had calmed, and the lake was as clear as a mirror with the trees reflecting off the water. Halley couldn't tell where the trees ended and the water began. A crow squawked, and a squirrel scampered in front of them racing up a pine tree.

The wilderness was so different than the crowded city of Houston. Out in the woods there were no traffic jams, no billboards or signs in the horizon, no sky scrapers, and no electronics. Halley loved her city, but going back to nature made her heart calmly lift like their puffs of breath in the cold air.

"Come this way." Ms. Mac interrupted her thoughts. "Today I am walking the campgrounds because there will be inspectors here tomorrow to evaluate the safety of the camp. We had several issues last summer, and they have been sent here by the National Summer Camp Board to make sure the camp complies with all the safety rules." Ms. Mac paused and turned to Halley, "Of course that is off the record and doesn't need to be mentioned in your article."

"Of course." Halley nodded. She had so many great memories of camp that she had almost forgotten the boulder that almost crushed her last year or when Gracelyn broke her arm in the Maze of Science. She had only written in her diary about how Ms. Spark tried to sabotage camp and had never mentioned it to her parents for fear they wouldn't let her come back.

"Tell me something, Halley. What was one of your favorite things about Camp Eureka last summer?" Ms. Mac continued walking down a path in the forest, side by side with Halley, using a giant walking stick.

"Oh, I loved being here with my friends and seeing all the wildlife." That magical moment when she and Gracelyn watched a deer standing in front of them at the edge of Lake Archimedes was one of her best memories of Camp Eureka.

Ms. Mac nodded and kept walking.

"Come to think of it, I liked seeing how much science is in nature already. Sometimes it's nice to get away and escape the expectations of school and the city." And get away from Mom. Was she talking too much? This interview was for Ms. Mac, not her.

A bright red color flitted in the woods, catching Halley's gaze. She blinked and looked closer. It was a cardinal, stark against the almost-colorless winter woods. It flitted about, almost following them on their walk deeper into the woods.

"Can you tell me more about this brochure? The map on it looks familiar." Halley inspected it closer noticing familiar features of the forest.

"That brochure you're holding is from the original camp that was here called Camp Sokatoa. It was a space and nature camp in the 1980s for kids who were interested in going into the space program. Believe it or not, I came here as a young camper and went on to be a teen counselor. When the time was right, I decided to run the camp myself along with a friend. We changed the name to Camp Eureka and changed the programs to a science and nature camp to give it a broader interest." Ms. Mac looked off in the distance with a twinkle in her eyes.

Did Ms. Mac want to be an astronaut when she was a kid? Who was the friend that Ms. Mac mentioned? Maybe it was Ms. Spark. "What was here before Camp Sokatoa?" Halley quickly changed the subject.

"Oh, I suppose the piney woods, but at one time there was an oil town not too far from here. Legend has it that early Texas settlers came here to seek their fortune in oil and minerals, and now their ghosts can still be seen at night trying to protect their treasures."

"Did you say ghosts?" Halley turned her head from the wilderness. Shivers ran down her back. Maybe they were being watched. They had made it to the Maze of Science, and Ms. Mac took her clipboard and opened the gates and closed them again, making sure they worked properly.

"Well, you can't have a fun summer camp without a good ghost story. It is all in good fun, I suppose." Ms. Mac turned to Halley and smiled. "Let's press on. We've got to make it around the entire camp before it gets too dark and we get too cold!"

Halley did not agree about ghosts being good fun. She could barely make it up Gracelyn's front porch with the witches cackling at her on Halloween.

"What treasures are these ghosts protecting?"

Halley blew on her hands to warm them up.

"Well, there is not only oil that was discovered here long ago, but there are rock formations and landforms that exist on this camp property that hide sought-after minerals."

"I love rocks and minerals!" Halley skipped along now as they were heading up a hill.

"Ah, and I believe you remember this rock formation." Ms. Mac stepped into a clearing. A large boulder as big as a car loomed in front of them and rested at the bottom of a hill with no sign of moving again.

"Yes, Unbalanced… I mean Balanced Rock." Halley stuttered. She was certain Ms. Spark had something to do with it rolling off its perch.

Ms. Mac feathered her fingers over the name of all the campers printed on the rock.

Halley searched for her name and found it. Halley Harper with a star flying out of the Y. And right next to it was Gracelyn's name with a rainbow coming out of hers. Halley's hand paused over Gracelyn's name, and she took a deep breath.

"What is the matter, dear?"

"Oh, it's Gracelyn. We are taking a break as friends, and our moms aren't speaking to each other."

"Is she still coming to camp next summer, dear? You used to be such good friends. What happened?"

"My science experiment made a big mess of their garage and could have hurt someone. I think Gracelyn's mom thinks I'm a disaster too. It's sort of a nickname that I've been stuck with." Halley's voice trailed. She wanted to talk more about the treasures around Camp Eureka. Not her science catastrophes.

"What is this nickname?" Ms. Mac stopped walking.

"Miss Disaster." Halley admitted.

Miss Mac's eyes softened. "Halley Harper, Miss Disaster? I like the ring to it." Ms. Mac chuckled. The wind picked up again and Halley could smell **ozone** in the air. "You seem more and more like me every day. Never forget your gift, Halley Harper, Miss Disaster. You've got the gift of turning ordinary into extraordinary. Now, let's head to the caves before this storm blows in. Then we'll head back to Energy Hall for some hot chocolate."

Halley felt light with the confidence Ms. Mac had in her. It was more than she ever had in herself.

Halley followed behind Ms. Mac and studied the old brochure map. The caves were east of Unbalanced Rock and someone had drawn a red X over the cave. Why did this map have an X over the caves? Did it mean X marks the spot? Could treasure still exist at camp? Her pulse quickened.

A gust of wind blew the map on itself, and a sprinkling of rain hit her face. She quickly folded it and tucked it back in her diary.

"I guess we need to cut our walk short, Halley. Let's hurry and head back to my office for some hot chocolate and marshmallows."

Halley was soaked from the rain when they finally got back to Energy Hall. Ms. Mac handed her a white Styrofoam cup with hot chocolate and extra marshmallows.

"Did you get all your questions answered?" Ms. Mac slurped on her own hot chocolate, "I appreciate you writing about Camp Eureka so that more campers will come."

"I think so." Halley assured her, brushing a crystal snowflake ornament with her fingertips. She knew Gracelyn would love one since they looked magical. Her insides were warm from the hot chocolate, and her mind danced thinking of magical snowflakes while she waited for Dad.

"You should talk to Gracelyn," Ms. Mac said. "Sometimes it takes a long time for friendships to form, and one slight bump will cause them to form a bit differently, just like these crystals. But in the end, they still make a beautiful structure to behold. My dad always said, 'In life you will be lucky if you find one true friend.' So if you think you found her, believe it will all work out for the best."

Ms. Mac eased a crystal snowflake from a branch and placed it in Halley's hand. "Here. Take this, and good luck with writing your article for the magazine."

"Thanks, Ms. Mac, thanks a lot." Did Ms. Mac have one true friend, and was that true friend Ms. Spark? Camp Eureka came close last year to shutting down for good. Where was Ms. Spark now? Had she and Ms. Mac ever made up?

"Say, whatever happened to—"

BEEP! Dad's car horn sounded, breaking her train of thought. "I guess I should be going. That's Dad, and he must be trying to get home before the rain gets worse. Thanks for the snowflake and the hot chocolate."

"Of course, dear. We'll see you next summer." Ms. Mac smiled and slurped on her hot chocolate. "Don't forget to grab next summer's flyer at the entrance."

Halley slowly walked down the hall and spotted a stack of colorful flyers at the door. She bent over to look at it more closely.

Camp Eureka
Friends and Experiments
Join us for chemistry, magic, and friendship!
Featuring the Greatest Magic Show
on Earth with Dr. **Alchemy**!
Come let the magic be at your fingertips!

Halley snagged the flyer. Maybe she could use it to write her article. More importantly, it could help convince Gracelyn to come back to camp next summer. Halley pushed open the door to Energy Hall, leaving behind the warm smells and memories, and ran toward her Dad's car in the gray, wet winter afternoon. She couldn't wait to give Gracelyn the crystal snowflake. Surely it would make up for what happened over Halloween.

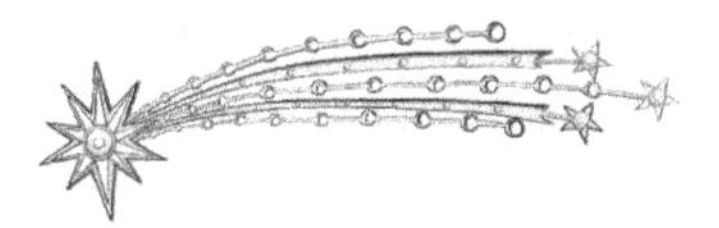

## CHAPTER 11
## A SNOWFLAKE CRYSTAL

HALLEY DIDN'T SAY MUCH TO Dad on the car drive home. She nestled the snowflake in her hand like a newborn chick. She didn't want to disturb a single crystal on the car drive back to Houston.

The rain pelted the highway, and sometimes the wind blew it sideways on the car. Dad was concentrating hard on getting home safely. The closer they got to their home in the city, tall office buildings replaced the forest trees and traffic jams replaced the wide-open spaces of camp. Halley already missed the calmness that she felt at Camp Eureka and couldn't wait for the summer to go back.

As soon as they pulled in the driveway, Halley pushed open the door and ran as fast as she could down the street towards Gracelyn's house, still cradling the beautiful crystal.

As she ran, all the things she wanted to tell her best friend flashed through her head. She wanted to tell her about seeing Ms. Mac, and how fun next summer's camp was going to be, and how she found a brochure with an X marking a treasure, and how sorry she was for Monster Toothpaste.

Surely this gift would help smooth things over with Gracelyn's mom and show her that Halley could be a kind, thoughtful, and responsible friend again.

The Halloween decorations covering Gracelyn's home had been replaced with a beautiful, sparkly red-and-green Christmas wreath. In the middle of the wreath a sign read, The Grinch Lives Here. Halley loved the Grinch. It reminded her that even the Grinch could be forgiven. Halley's heart lifted a little.

She carefully reached up and rang the doorbell. She held her breath. When she heard the lock turn, her stomach fluttered.

The door opened, and a wonderful blast of cinnamon aroma filled Halley's nostrils, like Christmas cookies baking. But instead of Gracelyn or Ms. Dee, a tall girl with long, blonde hair answered the door. She looked older than Halley and wore a cute red-and-green checked apron with a little bit of flour dusted over her cheek.

"Hi. Can I help you?" The girl smiled politely and obviously didn't know who Halley was.

"Um, hi. Is Gracelyn home?" Halley reached up and adjusted her comet pin holding her bangs in place. Who was this girl? Did she live here instead of Gracelyn? At that moment, a little breeze blew across the back of her neck and chilled her.

"Gracelyn? Oh, sorry she can't talk right now. She's busy. What was your name again? I'll let her know you stopped by."

"Yeah, tell her Halley came by to say hi and give her this." Halley held up the crystal snowflake, dangling it on her finger so it swayed in the breeze.

"Oh, that's so cute! Is that a crystal snowflake? Well, I'll make sure she gets it." The girl reached down and snatched up the snowflake before Halley could object.

"But…" Halley stammered.

"Thanks for stopping by. I'll let her know." The girl turned and quickly shut the door, making the wreath bounce up and down. The lock snapped into place.

Halley stood there stunned. Who was that girl who took her best friend's crystal? She rang the doorbell again and waited, but no one came to the door. That was weird.

She waited a little longer—stunned—before she turned and plodded back to her house.

A few raindrops landed on her face and thunder rolled in the distance. The storm must have made it to Houston. The clouds hung low in a lonely, dreary sort of way. She missed Gracelyn. She just had to figure out a way to get her best friend back.

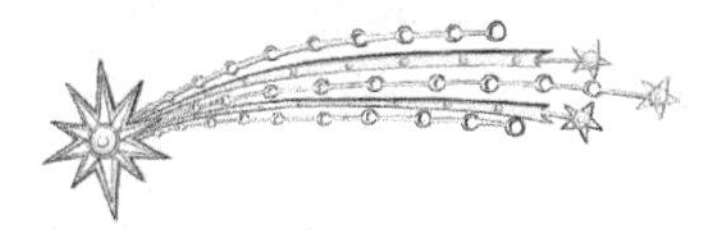

# CHAPTER 12
# VALENTINE

Dear Gracelyn,

*Happy Valentine's Day! How have you been? I can't believe that we haven't talked since Halloween. I thought you would want to see my published article, so I folded it and put it in this envelope. Please show it to your Mom, and maybe we can see each other again at Camp Eureka. Remember when you convinced me to go last summer? I hope you can make it. It just wouldn't be the same without you.*

*XOXO,*
*Halley*

*P.S. I found a treasure map and taped it to my diary so I don't lose it. Maybe you can help me find the treasure of Camp Eureka this summer!*

"Summer of Science: Why This Girl Is Heading Back to Camp Eureka" by Halley Edison Harper

I am so excited to be going back to Camp Eureka this summer! And you know what will make it even better—going to camp with old friends while making new ones!

Last year's science at Camp Eureka was learning about motion, but this year it's going to be about magic. I always thought that science could be even more magical when you are doing it with friends. There is going to be a special show with Dr. Alchemy, and the magic will literally be at our fingertips! I'm so excited!

We will not only do science magic, but we'll also learn about science in nature. I remember all the wildlife around Lake Archimedes and the giant rock that all the campers signed their names on. Now that our names are there, it's like we are part of the camp now.

The best part is there is still so much to learn about camp, like its caves, history, and treasures. Camp Eureka is so much more than just learning science. It's an adventure waiting to happen.

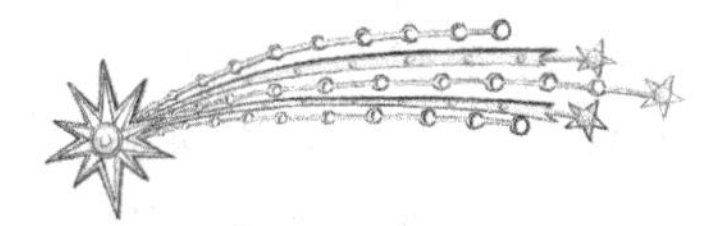

# CHAPTER 13
# A PIANO RECITAL

HALLEY PULLED BACK THE HEAVY, red velvet curtain. The bright lights made her cover her eyes as she stared out at the lone, black piano on the stage. Today was her piano recital. She tried to wiggle her toes, but they were squeezed in her tight dress shoes. She fidgeted in her dress and felt even more constricted and claustrophobic. A little shove on her back hurried her along on the stage to perform her piano piece. The clacking of her steps vibrated off the wooden stage floor and into the darkness of the audience.

Each step felt like trudging through Oobleck in slow motion and each step was harder and harder to take. She unlocked her gaze from the piano and looked out into the audience. All she could see were bits of dust floating in the air. Her bangs tickled her nose.

She brushed them aside and adjusted the pin to hold them back.

Someone cleared their throat in the crowd. Then a voice that sounded like Ben's yelled, "Go Halley!" Someone coughed before an awkward silence.

When she reached the piano bench, she pulled it out, and it made a low grating noise on the stage floor. She sat down, sat up straight, and crossed her feet underneath her. She gently placed her hands in starting position on the black and white keys. The keys were cold, just like her hands. She clasped her hands and popped her knuckles, nervously wiping her hands on her dress. She took a deep breath in and out, in and out.

Why did her dress feel so tight? It was getting hard for her to breathe.

She knew her piece by heart since she had been practicing it every day. She always loved "Fur Elise" by Beethoven. It was one of her favorites.

She placed her fingers in starting position again and started to press down on the keys, but they wouldn't move. No matter how hard she tried the keys wouldn't press down. They were stuck tight.

She turned her head to look for help. Perhaps the piano was broken? But instead of seeing her piano teacher, she saw Ms. Mac. What was she doing here? She blinked and turned around, looking into the audience. She tried to call to her Dad to come help, but no sound came out. Instead, a light on the front row illuminated Gracelyn and her long, blonde-headed friend with the crystal snowflake. They looked up and started laughing at Halley, pointing at her sitting there.

She looked backstage again, but instead of Ms. Mac, she saw Ms. Spark covered in pink gooey Oobleck pointing and laughing at Halley.

Halley tried to scream but nothing came out. She must be having a nightmare. She stood up from the piano to try and run but slipped and fell in pink Oobleck that covered the entire stage floor. She reached out to grab anything to hang onto, but she just kept falling and falling.

## CHAPTER 14
## IT'S CAMP TIME

"HALLEY, WAKE UP!"

Halley jolted and opened her eyes. She must have fallen asleep in the car with her head resting on the seatbelt. She rubbed her sore neck and squinted at the bright summer sun shining in the car window. Her head was groggy trying to make sense of the really weird dream she just had.

"Hey, are you alright there, Miss Disaster?" Dad turned his head to look at her. "You must have dozed off there and had some dream! We're almost at Camp Eureka. You might want to get your shoes on and get ready for your first day of camp!"

Halley reached for her shoes and brushed against the issue of *Empower* magazine laying open on the floor board to her published article. She scanned the title with her eyes, "Summer of Science: Why This

Girl Is Heading Back to Camp Eureka," by Halley Edison Harper. She made a silent wish that somehow Gracelyn read her article and would be at camp today. She closed the magazine, tossed it into her red backpack, and slipped on her red Converse tennis shoes.

Dad exited the highway and drove under a sign that read Camp Eureka. The energetic teen counselors were once again holding signs, dancing around, and waving to the campers. This time they were wearing colorful purple tie-dye shirts that read Chemistry Is Magic! Halley saw Cameron, her counselor friend from last year, wearing braided pigtails and purple knee-high socks. She was holding a bubbling beaker and started waving when she saw Halley and her dad pull into the parking lot.

Halley instantly spotted Ms. Dee's Volkswagen Beetle and her heart lifted. She would be at camp again with her best friend! Gracelyn popped out of the car and skipped to the open trunk to unload. Halley loved that Gracelyn had colored the tips of her blonde hair purple. She always wanted to try that. The other car door opened, and the tall, crystal-snowflake-thief girl stepped out. She also had colored the tips of her blonde hair. They looked like beautiful, blonde, purple-haired twins.

Halley reached up to touch her plain brown hair that had only grown out to a bob since Halloween. She always wanted to have colored-tipped hair, but Mom wouldn't let her. Halley's eyes prickled with tears. She just wanted to tell Dad to turn around and go home.

Dad leaned over. "You're going to have an amazing time here. Don't worry. You'll make a lot of new friends. And besides, you are a published author. You have a lot to be proud of."

"I just don't think I want to go anymore." Halley swallowed back the fast-food lunch that rose up from her stomach. Why was this girl with Gracelyn, and why was she at Camp Eureka? Camp was for her and Gracelyn, not anyone else.

"Remember what I told you last night," Dad suggested. "Try and say hi to Gracelyn, but make a few more friends. There will be a lot of new kids at camp that will be lonely to be away from home. Remember how you felt last year?"

"I remember, Dad." Halley decided she just felt a little carsick and needed some fresh air. She opened the car door and stepped out, grabbing her bags.

"Don't forget, Mom's going to pick you up at the end of the week because I'll be in Pittsburgh for work," Dad said. "And hey, I love you, Halley."

"I love you too, Dad." Halley smiled and took a deep breath of the piney forest air. She adjusted her half of her best friend necklace, shut the car door, and went to stand in the check-in line. She looked back and waved at her Dad.

"Oh my gosh! Halley Harper?" An excited voice squealed.

Halley turned around to see where the mousey voice came from.

"It is you! Did you know you are famous?" A girl shorter than Halley with glasses gushed. "Can you sign my camp flier? I saw your article in *Empower with Science* magazine, and I begged and begged my mom to let me go to Camp Eureka because I just knew that if I met you we would be best friends for life! Do you want to be my friend?"

"Um, sure." Halley looked around to see if anyone else heard this girl gushing over her.

"Ah! I can't believe I'm Halley Harper's friend." The girl hooked her arm through Halley's and snuggled up to her shoulder.

"What did you say your name was again?" Halley smiled and tried to twist out of her grip.

"My name is Charlotte. Oh, I just love your short hairdo. Maybe we can cut my hair to look like yours?"

Halley was speechless from the attention Charlotte was giving her and thankful that it was her turn to check in. She stepped toward the sign-in table, signed her name, and grabbed her cabin assignment. The counselor handed her a purple camp shirt and explained they would be tie-dying it for a camp activity later in the week.

Halley hoped that she and Gracelyn were in Cabin Curie together like last summer. She looked up and noticed a lot of girl campers had come this year. She smiled, her confidence building, thinking maybe her article influenced how many girls were at camp.

"Oh my gosh, what cabin are you in?" Charlotte said, interrupting her thoughts. Charlotte seemed to be good at butting in.

"Um, let's see, it says I'm in Cabin Anning. What about you?"

"Me too! Oh wow! I can't believe we are going to be in the same cabin! We're going to be best friends and roomies! This is awesome! Hey, it says here on the camp flyer that there is a boys' cabin called Cabin Avocado." Charlotte snorted. "I just love guacamole."

Halley squinted at Charlotte. Was this girl for

real? "I think it's Cabin Avogadro, you know, like the scientist?" Halley turned from Charlotte who was still laughing at her own joke and looked around for Gracelyn. She wanted to find out if she was in her cabin, and if not then she could always ask for a cabin transfer to be with Gracelyn. Then she could spend all week with Gracelyn. Maybe Gracelyn could help dye the tips of her hair purple too!

"So here's the camp schedule my mom went over with me before she left." Charlotte caught up to Halley who had started walking towards the cabins. "Today is Camp Kickoff with a Science Magic Show. Tuesday is Archimedes's Challenge. Wednesday is the Chemistry Race. Thursday is Glow-in-the-Dark Capture the Flag. Friday is the Science Talent Show and Parent Pick Up." With each day, Charlotte's voice grew louder and higher pitched.

Halley reached up to plug her ears and look for some way out of their conversation.

"Hey-ya Comet!"

She felt a gentle shove on her back and was thankful she was being rescued. She turned around. "Nathan! Oh wow, aren't you a sight for sore eyes!" She gave him a hug. Halley didn't mean to squeeze him so tight, but she was so glad to see a friendly familiar face.

"Hey, nice haircut! It's short, but I like it!" Nathan stared back at Halley. He was taller than Halley now and a bit skinnier than she remembered, but he had that same mischievous twinkle in his hazel eyes like he was up to something.

"Yeah, I like it short for the summer." Halley ran her hand through her hair and smiled.

"By the way, I read your article about Camp Eureka. Nice job! Did you know there is another science reporter for *Empower* magazine here at camp?"

"Really? Who?" Halley hadn't even thought about who the other kid science reporter was.

"She's right over there." Nathan pointed toward the camp check-in line. "There she is. Her name is Lexi Monroe."

"Who's Lexi Monroe?"

"Oh, I figured that you met her before." Nathan's eyes widened, and his voice got a pitch higher. "I helped her and Gracelyn unpack their bags from the car. Lexi is Gracelyn's cousin."

As soon as Nathan said her name, Lexi looked up and waved at Nathan. Nathan waved back then turned to look at Halley with a sheepish smile.

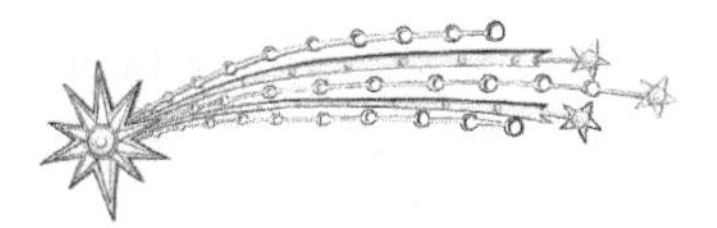

# CHAPTER 15
# THE SCIENCE MAGIC SHOW

HALLEY STARED AT NATHAN, WHO was half grinning now. "What?" He scratched behind his neck; his ears had turned a bright crimson color.

The newfound confidence she had a minute ago was gone again. Gracelyn must have a new best friend now. And what's worse? They were family! What was the old saying, "Blood is thicker than water"? Halley remembered Grammy telling her that family ties will always win over friends. Clearly a lesson in **density** that Halley didn't want to admit was true. She wondered why Gracelyn never told her about her beautiful, older cousin before.

Gracelyn waved back in Halley and Nathan's direction then looked over at Lexi. Lexi gave a disapproving glare in Halley's direction, grabbed Gracelyn's hand, and pulled her around so they weren't facing them anymore.

Wow. They used to be inseparable. Now they could barely say hi to each other. Had her mom and Ms. Dee talked since Halloween? Was friendship really that fragile even as an adult?

Halley came back from her thoughts and realized the last thing she wanted was to let Nathan know how much this Lexi girl was getting to her. "Excuse us, Nate." Halley turned on her heel and grabbed Charlotte's hand. "My friend Charlotte and I need to unpack our things before the Camp Kickoff." Halley adjusted her red backpack and looked down with an overly sweet smile at Charlotte. She secretly hoped Lexi and Gracelyn saw the two girls walking off.

"Okay, Halley, I'll see you around. I'll be in Cabin Avogadro this year." Nathan's voice trailed.

Halley turned back around to find Nathan still watching her and Charlotte before giving a wave and jogging toward some of the other boy campers.

Charlotte whispered, "Avocado, hee hee hee," and then gushed with the attention Halley was giving her for that brief moment. When Halley could no longer see Gracelyn and Lexi she let go of Charlotte's hand and crossed her arms. She had to formulate a plan to get Gracelyn away from Lexi.

"Are there really deer here? Do you think I could pet one? Did you know I love spiders? I have a pet spider that eats grapes. Tell me all about your favorite camp activity. What is your favorite camp food?" Charlotte held up the camp brochure like a tourist while they walked to their cabin. Halley felt like she was walking with a tourist from another planet.

Charlotte's questions kept her from planning

Operation Get-My-Best-Friend-Back. The only other sound that Halley could hear over Charlotte's voice was the crunching of pine needles as they walked up to the cabins. Why did Cabin Anning seem so far away? She looked in the direction of Cabin Curie and saw Gracelyn and Lexi opening the screen door.

"Here we are." Halley interrupted Charlotte's chatter and read the sign outside their cabin: Mary Anning, fossil hunter. "She sells sea shells by the seashore."

Charlotte read the sign aloud, "She shells she shells by the she-shore. I never could get this one right."

Halley rolled her eyes, shook her head, and tucked her hair behind her ears. She noticed the other girl campers had claimed the top bunks, so she tossed her red backpack carelessly on one of the bottom bunks that was left.

Energy Hall was buzzing with excitement after dinner. Halley tried to lose Charlotte by sitting with Nathan at dinner, but Charlotte was practically her shadow and rambled on about her favorite colors, favorite foods, and her pet rock collection. Halley stopped trying to answer her questions and ate in silence. The one thing that made Halley happy was there were a lot of girl campers this year and she was sure Ms. Mac was happy that camp seemed to be thriving again this year.

"Please take your seats, Campers. Kickoff is about to begin." Ms. Mac tapped the microphone several times before the campers followed her direction. "Welcome to the eleventh annual session of Camp Eureka!" Ms. Mac sang out. The stage behind her was set up like a science lab with colorful beakers, test tubes, and rows of soda bottles. "This is probably my favorite program we have in store for you tonight. We will fire up our imaginations by watching our very own Science Magic show!

"But first there are a few announcements; I'd like to ask my teen counselors to stand up so the campers will know who you are. I am the only director this year, and these counselors have kept things running as smoothly as they can. So thank you. Let's give them all a hand." Ms. Mac tucked the microphone under her arm and clapped loudly.

The campers clapped, and Halley started daydreaming that one day she wanted to be standing on stage with Ms. Mac as a teen counselor.

"Next, I would like to stay how nice it is to have more girl campers this year." Ms. Mac voice bubbled as she surveyed the room. "It had trailed off quite a bit in the last few years, but I'm glad you are back, ladies, to represent your amazing abilities in science!"

Charlotte elbowed Halley and whispered, "It's because of you." Then she put her head on Halley's shoulder. Halley couldn't help but smile and reached around to give herself a pat on the back.

"So with that," Ms. Mac lowered her voice, "I want to ask, can anyone tell me the name of a female scientist? I'm certain everyone knows a male scientist, like Albert Einstein, but what about a female scientist?" Ms. Mac scanned the crowd looking for answers.

That was too easy! Halley smirked, raising her hand first. She had stayed in Cabin Marie Curie last year. There were tons of female scientists that she could name: Jane Goodall, Dr. Temple Grandin, Mary Anning, the list went on and on.

"Let me see," Ms. Mac hesitated. "Someone new, perhaps." She called on Lexi, who was sitting up straight, smiling. Then Ms. Mac repeated, "Tell me, who is a female scientist?"

Halley turned her head, waiting for the answer.

Then confidently, Lexi oozed one little brilliant word. "Me."

Halley blinked.

Ms. Mac smiled. "That is exactly the right answer I was looking for, Lexi. Thank you very much. Everyone, give Lexi Monroe a round of applause."

"That was a good answer." Halley said under her breath. Why didn't she think of that? That applause should be for her! Shouldn't Ms. Mac thank her for bringing more girls to this camp? Halley couldn't believe what was happening. Lexi was taking over. She was taking over her best friend and now she was taking over her Camp Eureka. Halley glared at Lexi and crossed her arms puffing.

"Now, on with the science magic show!" Ms. Mac chimed, unaware of Halley's anger. "Allow me to introduce our guest scientist, Dr. Alchemy! Let's give him a round of applause!"

A man wearing a magician's top hat and tuxedo appeared on stage from out of nowhere and twirled his black-and-red magician cape to show a giant antique book underneath. He gently placed the book open on the table. "Thank you, Ms. Mac, but let me reintroduce myself. I am the Amazing, Astounding Dr. Alchemy!" He held out his hand flat and slowly brought his other hand over with a lit match!

Whoosh! A flame of fire flashed and quickly extinguished right in the palm of his hand!

"Whoa!" The campers said together and burst into applause.

"Thank you, thank you, Campers!" Dr. Alchemy dusted off his hands and paced the stage reaching to rub his pointy nose. "I was introduced as a scientist, but really I see myself more as a magician."

Charlotte looked over at Halley with a wide-eyed look and grinned.

"You see"—Dr. Alchemy stopped pacing—"Science seems stale like an old textbook, but magic is memorable! My favorite kind of magic involves the study of chemistry which is how materials behave when put in different conditions."

Halley was glued to Dr. Alchemy's every word. She couldn't wait to see what came next and wanted to figure out the science behind the magic.

"Now, for my next trick, behold this flask that appears to be full of water." Dr. Alchemy grabbed a flask full of clear liquid from the table. His eyes scanned the giant magic book, and he looked up, admiring the liquid he was swirling around. "If I add a bit of red liquid to this flask, what is going to happen?"

"It's going to turn the water red!" A few campers shouted out.

"Aren't you clever?" Dr. Alchemy grabbed another beaker full of red liquid and poured it slowly into the swirling clear water. "But let's see if you are right!" The liquid turned red as expected, then Dr. Alchemy shouted, "Presto-chango!" and suddenly the red liquid turned clear again.

The campers went wild with applause.

"Magic and science can involve color change!" Dr. Alchemy's eyes widened. "Yes, let me hear you clap for the Amazing, Astounding Dr. Alchemy!" And he held up the beaker for a few lucky campers in the front row to inspect closely.

Wow. Surely that wasn't water, but what was it? And more interestingly, what else was in that giant magic book of Dr. Alchemy's? Halley imagined all the magic tricks and their secret science solutions that were in there. Ben would love for her to do a science magic show for him when she got back from camp. If only she could just take a peek at what magic spells were in Dr. Alchemy's book!

"And now, ladies and gentlemen, I would like a volunteer to assist with my next magic trick!" Dr. Alchemy placed his hands on his hips and scanned the crowd.

Halley's hand shot up instinctively along with most of the other campers. Perhaps if Halley was called on, she could get a look at the magic book when she was up there. Charlotte sprang to her feet and started jumping up and down.

"Yes, you, you there who can't contain herself." Dr. Alchemy called on Charlotte in a silvery voice.

"Squeee!" Charlotte bounded towards the stage while a collective "aww" came from the other campers.

"I would like for you to stand right here in the

middle of the stage," Dr. Alchemy instructed. He was holding something under his cape, but Halley couldn't make out what it was. Charlotte still couldn't contain her excitement and was bouncing up and down.

"I sure hope you aren't afraid of a little magic bag of water." Dr. Alchemy grinned, revealing a clear plastic bag of water and a sharp pencil. He held the bag over Charlotte's head and the other campers giggled. Charlotte looked up and winced as Dr. Alchemy slowly started pushing the sharp pencil through the bag. The campers gasped, waiting for Charlotte to get soaked. "Soak her, soak her," some of the boys started chanting.

"KAZAM!" Dr. Alchemy shouted and pushed the rest of the pencil through the bag but water did not leak out. The pencil punctured both sides of the plastic bag like a skewer.

Charlotte smiled and touched her hair. "I'm not wet." She giggled and started to leave the stage. "Oh, but we aren't done!" Dr. Alchemy quickly produced a handful of sharpened pencils and one by one started skewering the pencils through the magic bag. "Kazam, kazam, kazam!" He shouted as he impaled the bag with the pencils one by one. Charlotte covered her head and squinted her eyes shut.

When he was done, Dr. Alchemy held up the magic bag impaled with ten pencils and not one drop of water came out. "Magic and science involves things that can be broken and magically come back together! Now, let's give my assistant a hand, shall we?" Dr. Alchemy grinned like he had secret knowledge and wasn't going to tell. The campers clapped wildly while

Charlotte jumped off the stage dry as a whistle.

Halley was still in shock. How did he do that? Surely science could explain the magic bag!

"And now the moment I've been waiting for." Dr. Alchemy grinned, turning and twirling his cape as he marched towards the rows of two-liter soda bottles. "If you study the magic of chemistry long enough, you will know that a **reaction** is when something is taken from a higher energy back down to a lower energy. Who here wants me to release the energy in these soda bottles?"

"Do it, do it!" The campers chanted.

"And just how are we going to do this?" Dr. Alchemy asked, unveiling a roll of white, hard candy and positioned them over the opened bottles of soda. "With magic, of course!"

Charlotte looked over at Halley and mouthed, "Are we about to get soaked?"

Halley winced. She just hated surprise noises and surprise food explosions.

"I hope you like soda!" Dr. Alchemy boomed before starting his countdown, "4, 3, 2, 1. Abracadabra!"

Halley closed her eyes to brace herself and looked up in time to see a row of soda geysers erupting one right after the other. Sticky soda rained down all over the campers who immediately burst into applause. Nathan and his cabinmates stood in the soda rain, trying to catch it on their tongues.

Gross. Halley covered her head avoiding most of the soda raining down on the campers in the front rows.

Dr. Alchemy took off his top hat and bowed, grinning from ear to ear.

"Well done, Dr. Alchemy." Ms. Mac shouted over the noisy chaos. "You really gave us a bang for our buck!"

The campers applauded, giving Dr. Alchemy a sticky standing ovation before they piled out the door of Energy Hall.

Halley stood amazed and clapped. It really was the greatest magic show she'd ever seen but she was not quite sure if it was magic or science. What was in Dr. Alchemy's magic book? She was going to have to find a way to take a peek for herself and prove it was all just science.

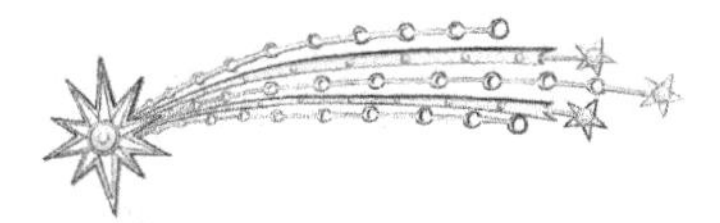

## CHAPTER 16
## THE MAGIC WORD

COME ON, HALLEY, LET'S GO back to our cabin." Charlotte insisted. "I can't wait to stay up all night just like a slumber party!"

Halley ignored Charlotte and instead watched as Ms. Mac and Dr. Alchemy talked to each other on the other side of the room. Her eyes darted back to the stage and strained to make sure that she spotted Dr. Alchemy's book lying there. Now was her only chance to look at it before anyone noticed. She just had to make sure she didn't slip in the erupted soda that was the aftermath of the show.

"Charlotte, can you wait outside while I go to the bathroom?" Halley fibbed, looking for a way to get space from Charlotte.

"Oh, I'll go with you! That's what best friends are for!" Charlotte gave Halley a side squeeze.

"No, really, just wait for me outside," Halley said, more sternly than she meant.

"Oh, okay, Halley," Charlotte turned around, shoulders slumped. "But let's walk back to our cabin together, okay? I'm afraid of the dark."

"Sure sure…" Halley motioned for Charlotte to leave and pretended to walk towards the bathrooms. Instead, she ducked behind the stage and spotted the book up close.

It had bits and pieces of paper tucked inside it and looked like her mom's cookbook stuffed with recipes. She pulled herself up on her knees and peered into the book, her nose grazing the inside spine. She had to look at it quick before the lights went out and before she got caught.

She had only really seen *Empower* magazine's experiments and some of Grammy's old science books. She swallowed. Her mouth was salivating with excitement. This book seemed different somehow from a dusty science textbook. It was magical, and she was getting closer to it, reaching out to turn one of its pages.

Suddenly, as if on its own, the book slammed shut.

Halley gasped.

"You forgot to say the magic word, Miss Harper," Dr. Alchemy sneered with his face so close to Halley's she could smell the peppermint on his breath.

"Um, I was just curious." Halley stammered, surprised Dr. Alchemy knew her name. "I just wanted to see what was in your book."

"You know a true magician never reveals his secrets." Dr. Alchemy grabbed the book, tucked it under his arm, and turned on his heel to leave.

"Just tell me one thing," Halley gulped. "There is science behind all of that magic, right?"

"Well, I guess that is for me to know and for you to find out." Dr. Alchemy gave Halley a toothy grin and sprung offstage. His cape made him look like a bat for a moment, flying off into the shadows of Energy Hall.

Halley's face and neck burned hot. She looked around. Had anyone noticed? Halley hopped off the stage and went out the door where Charlotte was sitting cross-legged on the ground. Nathan was leaning up against the wall. It was getting dark, and the last bits of sun in the sky made it look purple and crimson.

"Oh hey, Halley, here I am!" Charlotte gushed, jumping to her feet.

"What did Dr. Alchemy say to you?" Nathan pushed off the wall. "Can you believe that soda explosion? That was awesome!"

"I just wanted to see what was in his book. Why didn't he tell us the science behind the experiments? We are at a science camp, not a magic camp." Halley crossed her arms.

"Can you believe that I didn't get soaked with the magic bag full of water?" Charlotte hadn't listened to a word Halley just said. "I must have magic too!"

Halley looked over at Charlotte then back at Nathan, rolling her eyes as they started walking toward the cabins.

"That wasn't a magic bag, you know," Nathan whispered. "It was just the little molecules in the plastic sealing off the hole and holding the water in." Halley was glad Nathan was walking them back to their cabin.

"So, what happened with you and Gracelyn? Why

aren't you hanging out with her?" Nathan kicked the ground.

"I don't know. It's a long story but our moms got into a fight, and we haven't talked in a long time." Halley looked off into the darkening sky. "I've been trying to talk to her, but Lexi seems to be keeping her from me."

"Lexi? Lexi Monroe?" Nathan said, a bit starry eyed. "No, Lexi wouldn't do that. She is really nice. Don't worry. I'll get you and Gracelyn to be friends again." Nathan punctuated his statement with a firm nod of his head.

"Thanks, but it's complicated, Nate." Halley looked off as they passed Cabin Curie.

"Okay, well, I'll see you tomorrow for the first camp challenge." Nathan turned toward his cabin. "We must not be playing in teams this year since we were not assigned one. Too bad Team Comet has been disbanded this summer. It was fun winning challenges with you last summer."

"Goodnight." Halley waved at Nathan. It had been a long day, and she hoped it wouldn't be a long night. She really hoped Charlotte would just go to bed and not keep her up talking.

June bugs were flying around the porch light of Cabin Anning, excitedly bumping into each other. She could even hear their shells clicking together the closer they got. The rest of the cabin was unusually dark and quiet. What were the rest of her cabinmates doing?

She yawned and trudged up the porch steps. When she reached the cabin door, she stopped when she heard a muffled crash from inside the cabin, and she jumped back.

"Hey, did you hear that?" Halley's heart pounded in her chest.

"Hear what?" Charlotte asked and jumped behind Halley, using her like a shield, and peered over her shoulder.

"Back up," Halley said, almost running over Charlotte.

What had she heard? Maybe it was just the June bugs. Or maybe her cabinmates were already in the cabin with the lights out.

Halley took a deep breath and quickly swung open the cabin door, acting braver than she felt.

She reached her hand in to find the light switch and turned it on.

The cabin had been ransacked. Halley's backpack had been pilfered through and clothes and sleeping bags were strewn all over the floor. It looked like they had been robbed.

Then she spotted movement on the floor. A small animal was looking at Halley frozen and holding two marshmallows in each hand. It looked like it had been caught stealing cookies from a cookie jar.

"Ah, look, it's a cute, black kitty!" Charlotte marveled, releasing Halley's shoulders. "Come here, kitty!"

"That's not a cat." Halley warned, noticing the long, white stripe down its tail. She slowly backed up outside the cabin yanking on Charlotte's hand.

"Look out!" Halley's scream pierced the warm night air.

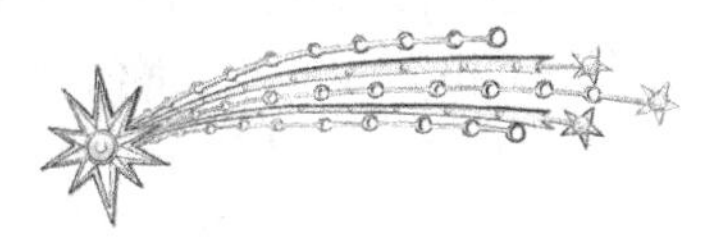

# CHAPTER 17
# MARSHMALLOW INTRUDER

"SKUNK!" HALLEY AND CHARLOTTE TUMBLED out of the cabin. The door slammed behind them leaving the creature inside eating his marshmallows. That was close. Halley recalled how bad skunks smelled when she passed squished ones on the road. How did it get in their cabin?

"Are you okay?" Nathan panted as he ran over. "What happened? You both look like you've seen a ghost."

"There is a skunk in our cabin eating marshmallows!" Charlotte stuttered.

"Well, that's why we don't bring food into the cabins. Food attracts animals out here in the woods." Nathan put his hands on his hips. "Now, calm down. How are we going to get him out of there? You don't want him to spray his stink inside your cabin!"

Suddenly, the ticking stink-bomb skunk poked his nose out from a hole in one of the window screens. It sniffed the air and then pushed its furry body out the window, plopped down on the ground, its tail high in the air. It turned around and stared at them.

"Oh look, it's going to do a trick!" Charlotte said as it stood on its front legs and arched its back at them. "Look, he can do a handstand like a gymnast!"

Halley's eyes started watering even before the smell hit her nose as the skunk loped off into the dark woods.

"Oh no!" Nathan held his nose and wrinkled up his face.

It didn't take long for the awful rotten-egg smell to reach the nearby cabins. Most of the campers came outside and started laughing at them and holding their noses. The other girls from Cabin Anning, along with Lexi and Gracelyn, walked up just after the stink filled the air. Ms. Mac appeared out of the darkness and pushed her way through the laughing campers.

All Halley wanted to do was disappear and get that awful smell out of her nose.

"Halley, did you have a skunk encounter?" Ms. Mac was trying not to laugh as she fanned her nose. "Campers, please stand back."

"It was in there," Halley whispered and motioned toward the cabin. Ms. Mac went in to inspect it and came out with an empty marshmallow bag dangling from her pencil. "Girls, who brought a bag of marshmallows into your cabin?" Ms. Mac stopped and looked between Halley and Charlotte.

"Um, I'm sorry." Charlotte hung her head. "I

just wanted to make s'mores with my new best friend, Halley Harper."

New best friend? Halley looked over at Gracelyn, who was fanning her nose next to Lexi. The s'mores Halley and Gracelyn had made in her microwave seemed like a distant memory. She was so mad at Charlotte right now.

"Campers, this is a good reminder to not bring food into your cabins. We are in the wooded wilderness! Nature is full of science, even science that stinks to high heaven. Luckily, the inside of your cabin didn't get sprayed, but you can't sleep in Cabin Anning tonight until the air clears. Hold your noses, grab your essentials, and let's head to another cabin. The show's over, everyone. Head back to your cabin and rest up for the first camp challenge tomorrow."

The only essential that Halley cared about grabbing was her diary. She shuddered. A wild animal had been in their cabin ransacking through their things!

Ms. Mac led them to Cabin Paschal, the teen counselors' cabin, and made space for them on the floor in spare sleeping sacks.

Halley got ready for bed and plopped down to inspect her diary. It was her only true friend here at camp other than Nathan.

"Gee, Halley, I'm really sorry about bringing the marshmallows. What are you looking at? Can I see?" Charlotte tried to sit next to Halley, peering over her shoulder.

"Charlotte, it's been a long day, and I'm just going to sit here by myself for a while and then go to bed, alright?" Halley snapped. She was annoyed that she wasn't sleeping in a comfortable cabin with her best friend, giggling about the skunk. Lexi was probably the one giggling. About Halley.

*Monday*
*Dear Diary,*

*The first night of camp stinks, literally.*
*My new cabinmate is annoying.*
*I want to go home.*

*Halley*

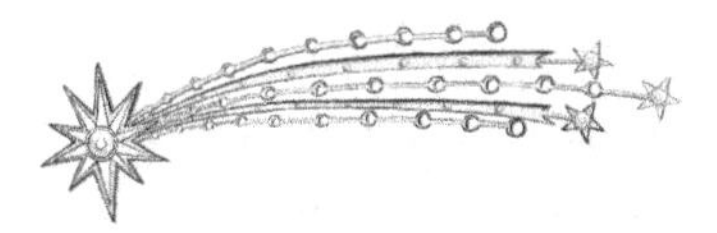

# CHAPTER 18

# EUREKA!

HALLEY OPENED HER EYES AND forgot for a moment what day it was. Why had she been sleeping on a cabin floor? Then she saw Charlotte brushing her teeth. Last night's memories of skunks, moving cabins, and listening to Charlotte snore through her nose filled her brain.

"Wakey, wakey, time for eggs and bakey!" Charlotte sang, plopping down on the floor next to Halley. "You were still asleep when the counselors and I went for breakfast, so I brought you back an egg-and-bacon taco!" Charlotte shoved a taco wrapped in aluminum foil at Halley.

"Thanks." Halley pushed herself up to sitting, rubbed her eyes, and reached for the taco.

"I figured you would need your energy today for the Archimedes Challenge! Do you think we are going to swim in the lake?" Charlotte pushed her glasses up on her nose and reached for her sunscreen.

"I'm not sure, but I guess that could be fun!" Halley loved to swim and loved to go fishing with her family back home. She finished the taco, balled up the aluminum foil, and quickly put her shoes on. She was thankful that getting ready was a lot faster now with short hair.

Halley stepped outside their cabin and breathed in the fresh morning air. The skunk smell was gone. Maybe today would be a better day. Maybe today would be the day she would talk to Gracelyn. Halley quickly walked towards the lake with a spring in her step.

"Halley… Hey, Halley, wait up!" Charlotte begged.

Charlotte reminded her of her tag-along brother Ben. Would she be able to go anywhere without Charlotte coming along? When they got closer to the lake, Halley spotted Gracelyn right away with her hair braided and chatting with Lexi. When she got closer, she reached out to touch Gracelyn's shoulder and then heard Lexi.

"You know, G, I thought of a great article I could write for *Empower* magazine," Lexi quipped, peering out of the corner of her eyes at Halley. "It would be titled, 'How to Use Science to Remove a Skunk Smell.'" Then she burst out laughing.

Gracelyn looked down, not amused by the joke.

Halley froze. Did Lexi actually say that? Was she making fun of Halley? If she was, why didn't Gracelyn defend her?

The only thing around here that stinks is you, Halley wanted to say but couldn't muster the courage.

"Come on, Charlotte." Halley grabbed Charlotte's hand but directed her comment to Lexi. "Let's get to the front so we can see better. I intend to win this challenge."

Halley held her chin high and pushed her way through the campers to stand at the front of the crowd. A large, white bathtub sat at the edge of the shoreline of Lake Archimedes. A few counselors stood around keeping the campers from getting too close. The starkness of the bathtub looked out of place from the rest of the surrounding nature.

Was the bathtub part of the challenge?

Nathan appeared next to her chewing on a blueberry muffin. "Want some?" He mumbled shoving a half-eaten muffin towards her.

"No, thanks." Halley grinned. Should she tell him what Lexi said or would he even believe her? She decided to keep it to herself.

A hush fell over the campers. Halley looked at the stark-white bathtub. Suddenly, someone who looked like an undersea diver in a black wetsuit jumped up from the tub, took off a snorkeling mask, and shouted, "Eureka!"

"Whoa!" Nathan stared, still munching on his breakfast muffin. "Campers!" Ms. Mac said dramatically, reaching into the bathtub and grabbing a submerged golden crown. "Eureka! I've found it!"

A few of the campers clapped, but most everyone was standing with their mouth open in awe of what just took place.

"How long do you think she was under the water snorkeling?" Halley whispered.

"That Ms. Mac is awesome!" Charlotte whispered back.

Halley smiled. Ms. Mac was quite the showman and knew how to get people's attention.

"Today we are going to explore density!" Ms. Mac announced, placing the crown on top of her head. She looked like she was the Queen of all Snorkelers.

"You may be wondering what a crown and a bathtub have to do with science. You see, our Lake Archimedes is named after a famous ancient Greek mathematician, Archimedes. Long before computers and calculators, he had to figure out if a king's crown was made of pure gold. This could have been easy to calculate if the crown was a simple cube, but it was an irregular shape.

"So after he drew his bathwater one night he noticed water spilled over the edge of the bathtub when he sat down in it. He guessed that if he collected this spill water then it would be the same space that he took up in the bathtub. This could be a way of calculating a volume of an irregular shape like a crown! He decided to drop the crown in the tub to find out its volume and calculate if it was in fact made of pure gold. Then he ran through the streets shouting, 'Eureka, I found it!' when he determined the density of the fake gold crown."

Ms. Mac motioned to the teen campers to help her out of the bathtub, and they grabbed a hold of her arms and helped her flop her flippered feet out. Halley grinned at Mc. Mac's theatrics. She wouldn't forget the image of her in the bathtub whenever she thought about density. Cameron walked up and gave Ms. Mac a small green watermelon.

"Now let's look at the density of this watermelon. Thank you, Cameron." Ms. Mac rolled the small melon around in her hands. "Do you think it is more or less dense than water? If it is more dense, it will sink. If it is less dense, it will float."

Ms. Mac slowly flopped behind the bathtub, still balancing the crown on her head, and looked over the tub before letting the watermelon roll out of her arms. The watermelon sloshed down—a bit of water spilled over the edge—and then it popped right back up, floating and spinning on top of the water.

"Wow!" The campers said and clapped. Halley had no idea that a watermelon would float. What other fruits float?

"Amazing, isn't it!?" Ms. Mac placed her hands on her hips. "Now onto the challenge. In the middle of Lake Archimedes are various floating fruits. You and one other camper will use a canoe to retrieve a fruit and bring it back to the shoreline. The first one back will be crowned for the day with your own Camp Crown of Archimedes! When you wear this crown, you can have anything in the General Store for free only for a day. I recommend trying the Rainbow Root Beer Float!"

Halley surveyed the red canoes floating gently at the edge of Lake Archimedes. Then she strained her eyes to see the floating fruits in the middle of the lake.

"On your mark. Get set. Go!"

Halley flinched and took off running to the nearest canoe, hoping that Charlotte would keep up with her. She wasn't going to let Lexi win the very first challenge, especially after that skunk comment.

Halley gingerly hopped inside the nearest canoe

and grabbed an oar while Charlotte bounded into the back causing them to almost flip over. “Careful, Charlotte!” Halley snapped and steadied the canoe.

“I can’t wait to be queen for the day!” Charlotte twittered, plopped down, and picked up the other oar.

Lexi and Gracelyn jumped in the canoe right next to them. Lexi’s eyes were fixed on the middle of the lake, and she was gripping the oar with white knuckles as Gracelyn sat down. “Okay, row!” Lexi shouted, and her and Gracelyn’s canoe took off towards the floating fruits.

“Hurry up!” Halley shouted at Charlotte and jabbed the oar into the water. “We’ve got to beat Lexi!”

“How does this oar work again?” Charlotte scratched her head. She seemed oblivious to the concept of a timed race.

Halley was just going to have to win this race by herself. She snatched the oar from Charlotte and attempted to row as fast as she could, but the harder she rowed the quicker the canoe turned in circles. Lexi and Gracelyn had left the bank minutes ago in a coordinated effort. Halley and Charlotte were stuck.

“Grrrr! This is so embarrassing.” Halley’s frustration made it even harder to row and she shoved the oar into the gooey muddy bottom of the lake to try and get back on course.

Then a hand steadied their canoe and gave it a gentle push towards the center of the lake. It was Ms. Mac. “Halley, work together with your partner and row.”

Halley smiled, grateful for the help. She conceded that she and Charlotte would have to row together even though it was slower than Halley wanted to go.

Several campers had already gotten to the middle of the lake, and Lexi and Gracelyn were working together to pull a watermelon into their canoe without tipping over.

Some of the boys had already capsized their canoe reaching for a fruit.

"Almost there!" Charlotte cheered and leaned towards a floating watermelon. Her fingers barely touched it and sent it spinning. She started to stand up. "Just a little closer!"

"Don't get up, Charlotte! Sit down!" Halley demanded and glanced over her shoulder to see Lexi and Gracelyn already gliding back to the shoreline with a watermelon.

Halley turned back and felt herself tip again. She used her oar like a tight-rope walker's pole to steady the canoe, but it was no use, her stomach lurched, and she knew they were going over in the water.

Then all she saw was brownish-greenish lake water.

She blew air from her nose and pushed herself up in the middle of the bobbing fruit.

"Help!" Charlotte called.

"Oh, here, Charlotte." Annoyed, Halley grabbed a watermelon for Charlotte to use as a floaty. Then she started pulling Charlotte and their watermelon to shore.

Lexi and Gracelyn had already gotten out of their canoe, holding up their watermelon. Ms. Mac crowned both Lexi and Gracelyn as the winners. The counselors hoisted Lexi and Gracelyn on top of their shoulders and carried them off towards Pie Are Square for lunch.

When Halley and Charlotte made it to the shoreline, Nathan helped them both out of the water.

"What happened out there?" Nathan's eyes crinkled in the corners holding in a laugh. "You know, we almost fell over in our canoe too."

"Thank you for your help, Nathan, but we got it just fine." Halley didn't want to be pitied. "Why don't you go and congratulate your friend, Lexi Monroe? We will win the next challenge, just you wait."

"You know, Halley..." All the laughter was gone from Nathan's voice, and he whispered, "You aren't always going to be number one."

Halley raised her chin high, glared at Nathan, and wrung out the bottom of her shirt. How dare he say that to her, especially after how mean Lexi was before the race?

"Lexi, Lexi... winner of the Archimedes Crown!" She heard the campers and counselors say in the distance as they went to lunch.

What was all the fuss about, anyway? It was just a dumb old crown.

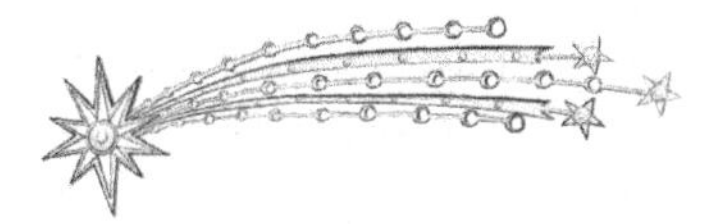

# CHAPTER 19

# A KETCHUP REVENGE

"YOU KNOW WHAT?" CHARLOTTE SAT munching away on her hamburger in the Pie Are Square dining hall. "My favorite color is brown because my beagle is brown."

Halley was trying to tune out Charlotte's chattiness as she stared over at Gracelyn and Lexi's table. Her head was throbbing from the sun, and her hair was still wet from capsizing in Lake Archimedes. She closed her eyes and could still see fruit bobbing in the lake water.

Lexi, still proudly wearing her Lake Archimedes crown, sat down with her food tray holding a hamburger and French fries. Nathan followed close behind her and sat down right next to her, close enough to brush her elbow.

Halley rolled her eyes. She should have been

nervous, but she wasn't. No one would expect a thing, and she was pretty sure no one had noticed when she took a half-empty bottle of ketchup, put a bit of baking soda in it, and carefully placed it back on Lexi's table.

It was a classic prank right out of *Empower with Science*. Halley wanted to try this one on Ben for a while, but desperate times called for desperate measures. Lexi took her best friend away and now she was taking away Camp Eureka. You know, Halley knew how to get a skunk smell out, by soaking in tomato juice. Well, ketchup would have to do.

Gracelyn joined Lexi and Nathan. Halley seethed. It used to be her sitting between the two of them, but now it was Lexi. She didn't feel bad at all for what she did and what was about to happen. Of course, she didn't want to have Gracelyn included in the prank. But at this point she didn't care.

Lexi picked up a french fry and held it up to inspect it. She turned to look at Nathan and laughed, giving him a shove with her arm.

What if Lexi didn't like ketchup? Halley panicked, knowing that Gracelyn always doused her french fries.

Then she saw Lexi ask Nathan to pass the ketchup.

Halley popped her knuckles nervously.

"Hey, Halley, what do you think about beagles wearing sweaters in the wintertime?" Charlotte broke into Halley's train of thought.

"Not now, Charlotte. I'm watching something." Halley nibbled on the edge of her chicken nugget. Was this going to work? She looked at the clock, only a few minutes now until lunch was over. Lexi tossed her head back and laughed, holding on to Nathan's arm.

Nathan blushed.

Lexi adjusted her crown, and slowly she reached for the ketchup bottle.

Halley took a deep breath.

Lexi shook the ketchup bottle.

Halley leaned forward.

GOOOOSHHH! A ketchup geyser erupted from the bottle and rained down all over Lexi, her crown, and her french fries.

Lexi jumped up and screamed. She looked down at her hands covered in ketchup and reached up to adjust her ketchup-covered crown.

"Bingo." Halley said under her breath, surprised it actually worked. "Just like *Empower* magazine said, a classic **acid-base reaction**."

Nathan grabbed the bottle, turned it over, and then spotted Halley staring at them. He mouthed the words, "Did you do this?" while pointing at the ketchup bottle. He must have remembered being on the receiving end of one of Halley's science pranks.

Halley ignored Nathan and thought her chicken nuggets became very interesting at that moment as she stuffed two in her mouth to finish lunch.

Lexi tried to fling the ketchup off her hands as she slopped away to clean up.

"Come on." Halley grabbed Charlotte's arm. "Lunch is over."

Dark clouds settled over Halley's heart as she and

Charlotte slowly walked back to Cabin Anning. Why didn't she feel justified pranking Lexi? The chicken nuggets sat in her stomach like rocks. Her legs felt like lead or one of the other heavy elements in the **Periodic Table.**

They opened their cabin, and Charlotte kept gushing about all the after-lunch camp activities.

Halley didn't feel like doing anything. She slumped into her bottom bunk bed and stared up at the underside of the top bunk wishing she could go home. She grabbed her diary and thumbed through a few pages, stopping to look at the old map of Camp Sokatoa.

"Come on, Halley. This is such a beautiful day outside. The sun is shining, the birds are chirping, and we could go make our tie-dye camp shirts or go play archery or go on a water slide." Charlotte kept chattering on and on.

"Go away, Charlotte. You are so annoying." Halley closed her diary shut.

Charlotte stared at Halley, stunned.

Halley wished she could take it back. "Oh, Charlotte. I'm sorry." She rubbed the back of her eyelids. "I didn't mean that."

"Oh, sure. That's okay, Halley." Charlotte turned and walked towards the door.

"No, wait, Charlotte. I'll go with you…" Halley's voice trailed off, and she propped her elbow up on her bunk.

"You know, Halley, I saw what you did back there in Pie Are Square. I saw that same prank in *Empower* magazine issue 14 myself. And you know, for as mean as you say Lexi is, you are kind of mean yourself." And with that, Charlotte turned, opened the cabin door, and ran towards the archery fields.

Halley swallowed. Charlotte was right. Halley wasn't being that great of a friend to Charlotte either. Maybe she shouldn't have come back Camp Eureka after all. Her best friend that she was trying to get back was only a few cabins away, but Halley couldn't help but feel all alone.

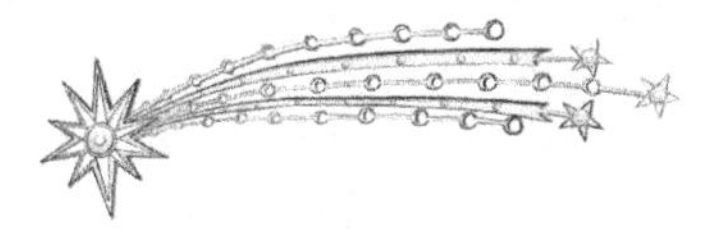

## CHAPTER 20
## ON THE TRAIL

HALLEY SHOVED HER DIARY UNDER her pillow and swung her feet from her bed. She decided to get some fresh air, thinking maybe getting out into nature would clear her head and make her feel better. She opened the cabin door and went onto the dusty trail behind Cabin Anning. The further she went back in the woods, the quieter the camp noises became.

She remembered back to Christmas when she and Ms. Mac walked these trails together. Everything was different now, but the woods felt familiar even though it was greener and hotter. It felt good to be alone. Suddenly she heard a rustling in the leaves. Maybe she wasn't alone. Maybe it was just a squirrel. Then the rustling got a little louder.

"Is that you, Halley Harper?"

Halley jumped and turned around.

Dr. Alchemy popped out of the woods onto the trail in his full tuxedo and magic hat. He was carrying his magic book and looked very out of place in the woods.

"Dr. Alchemy? I wasn't expecting to see you out here." Halley stared at the magic book, remembering how much she wanted to see inside the book. Maybe there was a magic friendship potion that would make everything better again.

"Oh sorry, I didn't mean to startle you. I'm walking up the trail to the neighboring camp. Want to walk with me a ways?"

"Sure." Halley forgot for a moment why she was even out there by herself.

"What are you doing out on the trail by your lonesome?" Dr. Alchemy shifted his book to the other side and walked next to Halley. "I thought you would be the most popular girl here at camp because of your amazing article in *Empower* magazine."

Popular. Halley felt the total opposite of popular. Even Charlotte was upset with her now. "Oh, I think I'm just going to hang out by myself. Not feeling too social here lately. By the way, how did you know about my article?"

Dr. Alchemy stopped and smiled. He opened the front cover of the magic book and removed a ripped-out magazine article. Halley noticed the title, "Summer of Science" by Halley Harper. "*Empower* is one of my favorite magazines of all time."

Halley smiled. "Mine too." Maybe magic and science were similar after all.

"This was my favorite line of your article." Dr. Alchemy cleared his throat. "The best part is there is still so much to learn about camp like its caves, history, and *treasures*. Camp Eureka is so much more than just learning science. It's an adventure waiting to happen."

The branches on a nearby tree swayed as if agreeing with Halley's written word. Halley took a deep breath, remembering when she wrote the article. She really wrote it as a letter to Gracelyn so she would come back to camp.

"You must really think it's special here." Dr. Alchemy interrupted Halley's thoughts as they kept on walking.

"Yes, I do."

"So much that you saved it from closing last summer?"

Halley stopped and turned to Dr. Alchemy. "How did you know that?"

"Oh, I've been talking a lot to Ms. Mac about camp."

"Well last summer was fun, but things have changed, I guess. To be honest, I'm just ready to go home now." Halley crossed her arms and stared down at her feet.

"I understand summer camps and how mean kids can be sometimes." Dr. Alchemy looked off in the distance. "Knowing you're not the best can be even harder."

Halley nodded, not wanting to admit he was right.

"Now I must run along to my next magic show up the road at another camp. Farewell, Halley Harper, and good luck here at Camp Eureka. I think your luck is about to change."

Dr. Alchemy tucked his book tightly under his arm, gave Halley a quick grin, and walked briskly away.

How odd that he was heading off into the woods to go to another camp. Ms. Mac never mentioned nearby camps within walking distance of Camp Eureka.

Maybe she should give camp another shot. It wasn't so bad, and besides, it was kind of nice having other girls here to talk to. Halley jogged back to her cabin with a renewed hope. She still had several days of camp left, and things could turn around. Maybe she would even get a chance to talk to Gracelyn and see her Archimedes crown.

When she reached the cabin, Halley decided to jot down a few ideas in her diary before heading to find the rest of the campers. She jumped on top of her bed and reached under her pillow, but her diary wasn't there. "Oh, where did I put it?" Halley looked around. Maybe it fell and somehow landed under her bed. She got down on her hands and knees and peeked under the bed. Nothing, just some dust bunnies.

"Okay, be calm. It's got to be here somewhere." Halley looked in her backpack. Maybe she tossed it in there. She even looked in the cabin closet, but all she found were a few old metal boxes with first-aid supplies. Her diary was nowhere to be found.

Her heart raced, and her face grew hot. What if someone was reading about her notes for the magazine or her inventions or even her thoughts about Gracelyn, Lexi, or Charlotte? Someone might be stealing her ideas of how to make a candy machine that Gracelyn had wished for.

"I've got to find my diary!" Halley panicked. What if Lexi was trying to get her back for the ketch-up explosion? How would Lexi even know where her diary was?

Suddenly the cabin door opened.

"Is this what you're looking for?" someone said.

Halley turned and saw Charlotte standing in the doorway.

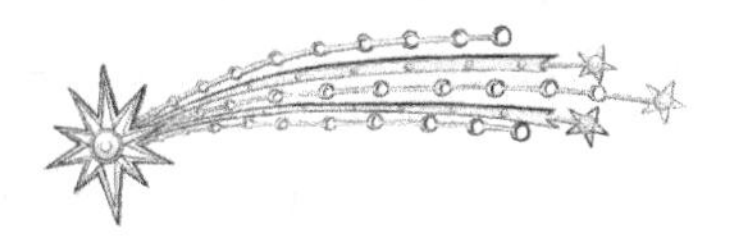

# CHAPTER 21
# GHOSTS

"OH MY GOODNESS! CHARLOTTE, YOU scared me." Halley exhaled the words, putting her hands over her heart.

Charlotte stood in the doorway holding up Halley's pen she used for her diary. How did she get that?

"This was outside the cabin. Isn't it yours?" Charlotte walked over to Halley.

"Oh yeah. That is my pen. Did you happen to see anything else outside?" If her pen was out there, maybe her diary was too.

"No, just this pen, but I'll help you find what you are looking for." Charlotte twisted her hands together. "I just wanted to say I'm sorry for annoying you earlier. I don't have a ton of friends at home and just wanted to feel special at camp."

Last year Halley had felt out of place until she came to Camp Eureka. Maybe Charlotte wasn't so different from her after all. Didn't they all just want to belong?

"No, I'm sorry." Halley took a deep breath. "I wasn't a great friend to you, and I haven't felt like myself since I've been here at camp. Do you mind coming with me to the Lost and Found? I think I may have lost my diary."

"Okay, sure. I'll walk with you." Charlotte turned towards the door. "But let's get back before dark. I heard some of the kids talking at archery about hearing ghosts around camp at night, and a few people said they could see figures glowing in the dark."

"Science can't prove that ghosts exist." Halley put her arm around Charlotte, giving her a side hug. As the girls walked out of the cabin together towards the Lost and Found room, Halley remembered that Ms. Mac mentioned stories of ghosts around camp. Maybe a ghost took her diary? Halley decided not to mention Ms. Mac's story to Charlotte. She seemed pretty shaken about what she heard at archery.

When they couldn't find her diary in the Lost and Found, Halley started to panic. She felt the color drain from her face, and her hands were clammy. She had to accept the fact that she wasn't going to find her diary that night.

What if Lexi took it? Would she use it against her after Halley pranked her with the ketchup explosion?

"Don't worry, Halley. Your diary will turn up soon." Charlotted offered, putting her arm around Halley. "Want to go to the campfire and make s'mores to get your mind off it?"

"Sure, Charlotte," Halley agreed. "That sounds like a great idea."

Charlotte was right. As soon as they walked up to the campfire and Halley smelled the sweet, gooey roasting marshmallows, she forgot all about her lost diary. She grabbed two marshmallows and sat down to roast one for her and one for Charlotte.

Then Halley spied Nathan was sitting around the campfire with a few other campers and was already in the middle of telling a ghost story. The faces of the other campers listening to him were ablaze from the fire. "Legend has it that Camp Eureka is haunted." Nathan slowly looked around at the campers sitting near the campfire. "It used to be an old oil town in the 1900s that had a busy railroad running right through it. After one hundred years, the railroad has been removed, but the road leading off into the dark forest still exists. Some kids say they've seen flickering lights along the road at night. Legend says that the ghosts of the oil workers have come back to claim their mineral rights and walk along the road at night searching for a camper to help them dig their treasures out of the ground." Nathan's voiced trailed.

"I thought I saw a ghost last night outside our cabin," one of the boys said.

"Maybe it thinks you'll dig up his treasure for him!" another girl quipped.

"Let's go, Halley. I'm scared," Charlotte whispered. "I don't want to hear any more stories."

"Want a marshmallow?" Halley offered, shoving the rest of her toasted marshmallow in her mouth. It was crunchy on the outside and gooey on the inside.

She turned on her flashlight, and hooked her arm in Charlotte's, and scanned the ground for the path back to Cabin Anning. Suddenly, her light caught a glimpse of red on the ground. "What is that?" Halley scanned her light again and ran ahead.

Charlotte raced to catch up with her. "Don't leave me behind, Halley," Charlotte stuttered. "It might be a ghost!"

There, nestled under a bush, was her diary. It looked like it had been carelessly tossed in the bushes. "Hey, it's my diary!" Halley shouted. "Why was it out here in the bushes?" She picked it up, brushed it off, and thumbed through it. A few of the pages stuck together and a little bit of the ink had smeared, but it was mostly okay. Then Halley noticed a page was missing. That's odd. It was the page she had taped her old brochure of Camp Sokatoa to. But why would just that page be missing? Now she was convinced that someone stole her diary.

She looked around for Lexi but didn't see anyone but Charlotte looking at her with wide, questioning eyes. She shuddered and tucked her diary under her arm. "Come on, Charlotte. Let's get back to our cabin."

"If it's all the same to you," Charlotte said, "I'm going to leave the bathroom light on tonight, okay?"

The girls ran back to Cabin Anning, and they both agreed to sleep right next to each other for safety.

Halley opened her diary, thankful to have her old friend back, and clicked on her flashlight to write.

*Tuesday*
*Dear Diary,*

*I wish you could talk. Where were you!? Who stole my old map of Camp Sokatoa? If I don't see it again I think I remember that there was an X over the caves. I wonder if there is really a treasure buried there?*

*I'm so glad you are back. Today we learned about density and got soaked in Lake Archimedes. Note to self: The ketchup explosion worked!*

*My new friend Charlotte reminds me a lot of how I felt last year at camp. I think we could end up being really good friends. I still haven't talked to Gracelyn. I miss her.*

*Are ghosts real? They can't be proven scientifically, can they? Maybe I'll sleep with my flashlight on tonight.*

*Love, Halley*

# CHAPTER 22
## AN ENCOUNTER

HALLEY OPENED HER EYES AND thought she heard something. She turned and saw it was just Charlotte snoring and the rest of her cabinmates still sleeping. She rubbed her eyes, gritty from last night's campfire. Her hair still smelled like s'mores, and she was still clutching her diary, but she rolled out of bed and stared out the window. The morning light was gray and peaceful, and she heard a few birds chirping a morning song. She walked out on the cabin porch and crossed her arms against a chill in the air. It was a welcome relief to the blazing sun they might see later in the day.

She walked toward the lake to get a better look at a low white cloud hovering right over the water. *That's odd.* It almost looked like little ghosts hovering over the water. She walked a little faster towards the lake, fascinated at the lake fog. The camp was quiet, and she decided she must be one of the first campers up that morning.

"There are no such things as ghosts," Halley said aloud as she stood quietly at the bank of the lake, trying to convince herself. She reached out her hand, trying to catch the tiny water droplets. "It's just **evaporation** or fog that happens when the warm air over the lake is cooled by the cold air above it."

"Hey, Halley, is that you?" a familiar voice said.

Halley turned. "Gracelyn? Oh my gosh, I'm so glad to see you! How are you?" Gracelyn was alone, and Halley ran over and gave her the biggest hug ever.

"I'm great." Gracelyn beamed after such a big hug and reached out to touch Halley's hair. "I love your new hairdo. I'm so glad you came to the lake, but I don't have much time to chat. I snuck away while Lexi was taking a shower. I figured I'd find you here checking out this fog. It's pretty magical, isn't it?"

"Oh my goodness, Gracelyn, camp just isn't the same without you," Halley gushed. "Sometimes I just feel like I shouldn't even have come back. Why is Lexi not letting us talk to each other?"

"My mom wanted her to go to camp with me to watch out for me. She didn't want me to get hurt like I did last year." Gracelyn rubbed the wrist that she had broken last summer at camp. "Lexi is super competitive, and she thinks that you are a competition to her, I guess."

"But why won't she even let me talk to you?" Halley swallowed against the tightening of her throat.

"I don't know. I've tried to contact you a few times

back at home, but I didn't want Mom to find out and get mad. I think our moms are in a big fight." Gracelyn looked off beyond the lake. "Hey, do you remember when we saw that deer here last year? That was so cool." Gracelyn changed the subject.

"I'll never forget that." Halley's eyes welled up with tears. "I'm so sorry for the Monster Toothpaste accident. I'm sorry I made your mom so mad. And I'm sorry for whatever my mom did. I miss you, Gracelyn, and I miss us being best friends."

"I know. I miss you too," Gracelyn admitted, and the two girls hugged. "I need to get back before Lexi suspects that I'm gone." Gracelyn peered over her shoulder. "Mom has been calling up to the camp every day to check on me and make sure everything is alright."

"I'm going to fix this." Halley promised. "I'm not sure when or how, but somehow I'm going to make things right again. Promise me that we will be best friends forever no matter what happens." She wanted to ask Gracelyn so bad if Lexi took her diary, but she was going to have to figure out some other way.

"Promise." Gracelyn looked at Halley with a tear forming in her eyes. She reached down and showed Halley the other half of their best friend necklace hidden away under her shirt. "I love you, Halley."

"I love you too." Halley wiped the corner of her eye. The two girls hugged, and then Halley watched Gracelyn turn and run towards Pie Are Square for breakfast.

Halley stood there watching her best friend and reached down to touch her half of her best friend necklace. A breeze started blowing and lifted the lake fog.

Suddenly, she saw something moving in the distance behind the boys' cabins. It was a man in a top hat and cape racing off into the woods holding a giant book of paper clippings tucked under his arm.

"Dr. Alchemy? Hey, wait up!" Halley called out, running to catch up with him.

He was in a hurry, struggling to keep all the paper clippings from falling out of his book and too far off to hear Halley. A few clippings fell on the ground, and a breeze tossed them in the air. Halley ran up to try and keep them from blowing away in the breeze. She stepped on a paper clipping with her foot, reached down, and realized it looked familiar. It was the old camp brochure taped onto one of her ripped-out diary pages!

What was he doing with a page from her diary? Did Dr. Alchemy steal it? All along she thought it was Lexi! Curiosity got the best of her now, and she kept following Dr. Alchemy deeper and deeper into the woods by hiding behind trees and bushes every few yards.

Dr. Alchemy was in such a big hurry, going deeper and deeper in the woods, that he had no idea she was following him. Eventually, he stopped at the opening of a giant, hollowed-out rock formation and looked around, rubbing his hands together.

Halley dove behind a fallen tree and cautiously peered over it. So this must be the caves that Ms. Mac was going to show her when it started raining! But why was Dr. Alchemy over here near the caves? He paced in front of the cave opening, looked down at his watch, and then put his hands on his hips with his lips pursed together.

Halley was out of breath from trying to keep up with him, and she tried to calm herself so her breathing wasn't so loud. It was so quiet out here in the woods that Halley had to be perfectly still or Dr. Alchemy might hear her. She looked down at the old camp brochure she was clutching in her sweaty hands and decided they were near the X that was marked like a treasure map. Suddenly, a twig snapped behind her. Dr. Alchemy looked up startled and began walking right toward the fallen tree where Halley was hiding!

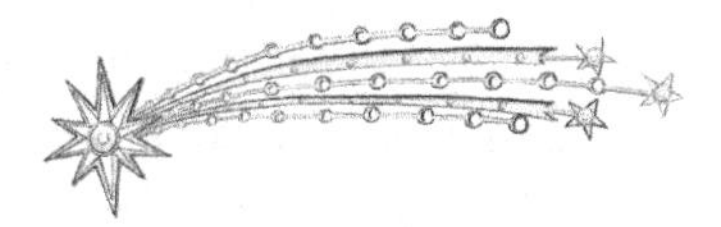

# CHAPTER 23
# THE CHEMISTRY RACE

HALLEY FROZE. WHAT WAS SHE going to do? She crouched in a little ball behind the fallen log and held her breath.

Then from somewhere behind her, a tall, skinny woman emerged from the woods. Halley recognized her instantly even without her holding her clipboard or covered in Oobleck. It was Ms. Spark.

"There you are!" Ms. Spark hissed, picking up her feet to step over the fallen log. She didn't even notice Halley and had a pinched expression, glaring at Dr. Alchemy.

"Hi, Aunt Spark." Dr. Alchemy squeaked, rubbing his nose and taking off his top hat to hold in front of him.

"Well, where is it?" Ms. Spark peered around the opening of the cave. "Where's the treasure?"

"Uh, I haven't found it yet." Dr. Alchemy stuttered. "It's a lot harder to find than I thought. That map wasn't the most precise, you know."

"Don't sass me, mister. We've got no time to waste." Ms. Spark pointed a long, bony finger at Dr. Alchemy. "Those little brats are at their challenge on the other side of the camp, and I've got to go back tonight. Let's check the caves again for any clues. It's got to be in there somewhere. If we don't find it today, you're going to be digging for it tomorrow!" Ms. Spark barked and plodded towards the opening of the cave. "I just hope that Halley Harper girl hasn't figured it out yet!"

Figured out what? What treasure was really in those caves? Why was Ms. Spark looking for it? Halley had to get back to camp to let Nathan and Gracelyn know that Ms. Spark was back and that Dr. Alchemy was her nephew. And what's worse, he stole her diary!

When they both ducked into the cave opening, Halley decided it was her chance to make a run for it. She turned and ran back in the direction they came, hoping that as long as she ran in a straight line south, she would eventually see the opening of camp. She just kept running and not looking back, jumping over branches and around trees. Her mind raced, her legs started to burn, and her lungs felt like they couldn't keep up with her breathing and her thoughts. She was running so fast that water streamed from her eyes. Halley blinked.

"Oof!" Halley practically fell when she tripped into the skinny arms of another camper.

"Halley?"

"Cameron! Thank goodness it's you!" Halley could barely form the words as her breath was coming too fast.

"Where have you been? We've been looking all over for you." Cameron stared at Halley. "The next challenge has already started, and we thought you might have overslept, but you weren't in your cabin."

"But I've got to tell someone what I saw." Halley panted, but Cameron didn't seem to understand what she said.

"I hope you're ready to run. The challenge is the Chemistry Race. I'll explain the rules when we get to the starting line." Cameron went ahead, her braided red ponytail swinging back and forth as she jogged.

Halley stopped and put her hands on her knees, trying to catch her breath. Great, the camp challenge was a race? She could barely catch her breath, and her mind was racing thinking about what Ms. Spark and Dr. Alchemy were up to. But she didn't want to wait there and have them find out she was eavesdropping on them. "Wait up for me, Cameron!"

By the time they got to the starting line, the rest of the campers had already raced ahead. She could see Lexi was leading the pack in first place, and Charlotte was walking with a few other campers in the last place.

The last thing Halley wanted to do was run a race after what she had just discovered. She really needed to catch up with Nathan and tell him what she saw in the woods, but Nathan must have been in the pack of campers running together. "So, what is the Chemistry Race?" Halley turned to Cameron.

"Okay, you need to hurry if you want to win, but basically the goal is to be the first to cross the finish line after decoding a secret message. Here is your secret message. Don't lose it."

Halley looked down at the blank piece of paper in her hand. She held it up to the sunlight. "Oooh, I love secret messages!" Halley smiled, forgetting for a moment about Ms. Spark. "What do I need to do to decode it?" Halley hoped Cameron would give her an inside scoop.

"Basically, this is an acid-base reaction. To solve the riddle, you need to find something acid in nature. I'll give you a hint, there are dark-purple wild grapes growing along the race route. Their juice is very acidic, so don't eat them. It will make your mouth itch!"

"Thanks for the hint, Cameron." Halley smiled and raced off to catch up with the other campers.

"Good luck, Halley!" Cameron shouted behind her.

Her stomach gurgled. Come to think of it, she forgot to eat breakfast. She spotted Charlotte and sprinted as fast as she could to catch up with her. Charlotte was walking, fanning herself from the steamy summer morning.

"Hey, Charlotte! Why aren't you running?" Halley panted.

"I've never liked running," Charlotte sighed and rubbed the sweat on her nose, making her glasses slip down. "Why even try? I'm not going to win."

"Now that doesn't sound like the Charlotte I know!" Halley cheered. "Come on, I want to catch up with Nathan to tell him something, and I know what we need to do to decode the secret message!"

A few vines heavy with wild grapes were growing

along the race route just like Cameron hinted. The other campers must have not figured out how to decode the message. Most of them were running as fast as they could to the finish line with blank papers.

"Start picking a few grapes, and we'll squeeze the juice over the message to decode it!" Halley's stomach gurgled again. "You know what was funny, Charlotte? I saw Dr. Alchemy in the woods before I got here. I'm not sure what he was up to, but he stopped in front of the caves and…" Halley looked over to see if Charlotte was listening.

"Hey, Charlotte, did you hear what I said?" Halley looked in Charlotte's direction and saw her popping a wild grape in her mouth and starting to chew. "Charlotte, wait! You aren't supposed to eat them!"

"Why not?" Charlotte said. "They look like they are from the grocery store, and besides, I'm hungry!"

"They are very acidic and might make your tongue itch!" Halley warned.

Suddenly, Charlotte spit out the grapes, reached up, and started trying to scratch her tongue. "Oh my goodness, my tongue is super itchy. Am I going to die, Halley?" Charlotte looked more and more panicked.

"You're going to be okay, but let me help you to the medic." Halley grabbed a few grapes and stuck them in her pockets. They retraced their steps to the starting line. Charlotte had calmed down when they noticed a few of the teen counselors talking with Cameron.

"Can you get Charlotte some water?" Halley asked. "She ate a few wild grapes, and she may need to go to the medic."

"I'll be fine, Halley. I don't really want to race

anyway." Charlotte wiped the sweat off her forehead and took a drink of water. "Really, you go ahead, and I'll wait for you at the finish line."

"Are you sure? There is no way I'm going to win the race now." Halley looked down at her blank paper. She was so curious as to what it said. She reached into her pocket and pulled out a grape. "I just wonder if this will work and what the message says." Halley laid the message flat on the ground, bent over it, and squeezed the dark purple juice over the paper.

And like magic, a message appeared.

UNLOCK THIS MESSAGE.
RUN TO THE FINISH LINE QUICK.
INSTANT COUNSELOR.

"Counselor?" Halley whispered. "You mean you can be a counselor if you win this race?" She looked over at Cameron.

Cameron smiled. "You can hang out with me and be a counselor for the day!"

Halley beamed. She would really have to run fast to win first place and hope that someone hadn't decoded the message before her.

"I'm going to win this." Halley looked at Charlotte and Cameron. She clutched the decoded message and took off running. Her mind raced as her feet pounded the dusty ground. If she was a counselor, she might be able to figure out what was going on with Dr. Alchemy and Ms. Spark. Maybe she could enlist the counselors' help.

She passed the wild grape vines and wondered where the rest of the campers were. After running for several minutes, she rounded the corner into an opening and saw someone with long braids sitting on the ground hunched over. As she got closer, she realized it was Lexi, and she was holding her ankle.

"Lexi, are you alright?" Halley jogged up and noticed that Lexi's deciphered message was next to her. "Where is Gracelyn?"

"I don't know. I sent her ahead to get Ms. Mac, but I haven't seen her in a while. I think I sprained my ankle." Lexi winced.

"Here, let me help you." Halley grabbed Lexi's hand to help her up.

"I can't walk. It hurts really bad." Lexi winced as she held on to her ankle. Tears smudged her reddened face. She must have been running hard before she fell. Lexi wiped her nose with the back of her hand.

Halley looked around. No one was coming behind her, and she couldn't see anyone ahead of her. Halley knew Lexi was the winner. She had deciphered the message and was clearly ahead of Halley before she had fallen. And besides, Lexi hadn't taken her diary, Dr. Alchemy had.

For a moment, Halley thought back to Ms. Mac giving her the crystal ornament. She thought back to her Grammy and the bobby pins for her hair. She even thought back to her mom who had gently cut her hair and assured her that everything was going to be alright.

"Come on, Lexi." Halley threw Lexi's arm over her shoulder and put her arm around Lexi's waist. "You're coming with me."

"What are you doing, Halley? Don't you want to beat me in the race? Why aren't you going on without me?"

"Because you are the winner, Lexi. Now come

on." Being a friend was more important than winning.

It was a struggle as she and Lexi moved slowly forward, clutching their decoded messages. Halley felt light-headed from not having breakfast, and her legs were burning from holding onto Lexi. After what seemed like forever, they rounded another corner and saw a sign that said Chemistry Race Finish Line. Many of the campers were sitting along the sidelines. She could hear Lexi wincing while she used Halley as a crutch.

"Come on, Lexi… not much further now."

"Okay, Halley…"

Halley had a razor-sharp view of the last few steps she needed to make. Every step seemed like it was in slow motion, her feet plodding as they neared the finish line. It wasn't about being a counselor for the day. It wasn't about a silly crown. It was about doing the right thing.

Ms. Mac ran up to the finish line with Gracelyn close behind her. Halley stopped just short of the finish line and carefully heaved Lexi into Ms. Mac's arms on the other side. Lexi won the race. Halley bent over to catch her breath and heard loud cheering from the counselors and campers who just witnessed what happened.

*Wednesday*
*Dear Diary,*

*Today I learned four things.*

- *Don't eat wild grapes.*
- *Dr. Alchemy took my diary, and Ms. Spark is his AUNT!*

- *There is a real buried treasure at Camp Eureka!!!*
- *Doing the right thing feels better than winning.*

*Love,*
*Halley*

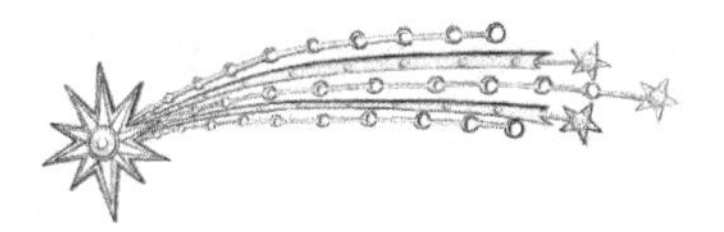

# CHAPTER 24
# THE TREASURE OF CAMP EUREKA

"DID I HEAR YOU RIGHT, Halley?" Nathan leaned closer to her, pushing up the safety goggles on his face. "Did you say there really is a treasure at Camp Eureka?"

A flash of light and clap of thunder startled Halley, and she jumped, looking out the window of the SmArt craft room. Rain started pelting the roof of the building. Halley was so glad she was inside with the old Team Comet making their tie-dye camp shirts.

"SHHH!" Halley and Gracelyn whispered together across the table at Nathan. "We don't want everyone to find out!" Halley looked around to see if anyone heard them, but the rain made it hard to hear anything. She quickly changed the subject when a few campers looked in their direction.

"How's Lexi's ankle?" Halley cleared her throat. It

felt so good to be chatting with Gracelyn again.

"She's going to be okay. The medic told her to rest and stay off of her ankle. She took a pretty nasty fall during the race. It was so nice of Charlotte to offer to keep her company by playing dominos with her. In fact, I think that Lexi has a new best friend!" Gracelyn looked at Halley and winked.

"I'm glad to have mine back." Halley hugged Gracelyn. "At least while we are here at camp."

"Okay, enough with the sappy stuff." Nathan plopped down on a chair beside Halley and scooted closer. "Let's get back to what you saw yesterday at the caves. Did you really see Ms. Spark again?"

The three friends huddled close together. "Well, I started following Dr. Alchemy when I realized he had dropped a page of my diary with an old treasure map that I found here at camp. Then Ms. Spark walked up and started talking about a treasure. I think I'm more creeped out that Dr. Alchemy went through my diary! Eww!"

"That Dr. Alchemy always seemed a little kooky." Nathan held up his tie-dyed shirt. "Do you like?"

Halley nodded and dipped her twisted T-shirt into the hydrogen peroxide.

"Do you think that's his real name? You know what alchemy means, don't you?" Nathan whispered. "It is trying to turn ordinary things into gold, you know, like magic."

Halley shivered.

"I say, let's get to the treasure before Dr. Alchemy does!" Nathan grinned, looking at the girls over the top of his safety glasses. He had a twinkle of adventure in his eyes.

“What if it’s too late?” Halley questioned.

“But what if it’s not?” Nathan leaned in. “Come on, let’s go on a treasure hunt!”

“Do you think we’ll run into any ghosts?” Gracelyn whispered.

“We may not run into a ghost but we could run into Dr. Alchemy,” Halley warned. “We have to be careful and make sure no one knows we’re missing from the camp activities.”

“Tonight is Glow-in-the-Dark Capture the Flag. No one will notice that we’ve snuck away to the caves.” Nathan plotted.

“What if they already found the treasure?” Gracelyn looked around. “What would we do then?”

“I’ve got an idea.” Halley tucked her hair behind her ears and inspected her tie-dyed Camp Eureka shirt. “Nathan, can you grab a shovel, and Gracelyn can you find a few flashlights? I’ll grab the treasure map and an old first aid box I found in my cabin. There are a few things I want to borrow from the SmArt craft supplies.” She reached over and grabbed the hydrogen peroxide they had used to dye their shirts. “I don’t think Ms. Mac will mind if it’s used to save the Treasure of Camp Eureka.”

Halley took a deep breath of the cold, damp air that hung around her, and she blinked her eyes, trying to adjust to the darkness. Something fluttered by her cheek, and she ducked, hoping beyond hope that it wasn’t what she thought it was… a bat.

Her arm prickled with goose bumps.

She looked down at the shovel in her hands and was thankful for the flashlight that was propped against the cave wall where she had stopped to dig.

*I hope this works.* She put the shovel down, lifted the metal box from the hole, and brushed off the dirt. The treasure of Camp Eureka. Was this what the ghosts had been looking for?

She hoped now that what she felt was a bat and not a ghost! She hoped the flashlight didn't lose power and before she made it to the opening of the cave. The light was the only thing saving her from feeling claustrophobic. She had to get out of the cave quickly.

"Halley Harper." She suddenly heard a familiar voice. Halley spun around expecting to see someone else. She tried to hide the box behind her back as best she could.

"Dr. Alchemy!" Halley said, braver than she felt. "What are you doing here?" Did he always wear his magician's hat and cape?

"I was going to ask you the same question." Dr. Alchemy shone his flashlight in Halley's eyes. "Shouldn't you be playing Capture the Flag with the other campers? You don't want to miss out on the fun."

"Oh, I suppose I should get going." Halley squinted towards her own flashlight. How was she going to grab it without dropping the treasure box? More importantly, how was she going to get out of the cave with the box while Dr. Alchemy was standing in her way?

"What do you have behind your back, Halley?" Dr. Alchemy pursed his lips together. "Do you need my help to carry it?"

"Um, help? Uh no. I've got it. It's just a bit of a camp... project." Halley stuttered, her gaze darting in the direction of the cave entrance. Her heart pounded. What if he didn't let her out of the cave with the box?

"Hand it over, Harper." Dr. Alchemy didn't seem so nice now. "We both know what you've got, and I'm not about to let you take it."

Halley took a deep breath. There was no turning back now. Only one of them would leave with the treasure.

"This box? I'm sorry, no. You can't have it, and besides, scientists never reveal their secrets." Halley said bravely, remembering back to when she wanted to see Dr. Alchemy's magic book.

"You have no idea what you've got in that box, and I need you to hand it over so no one gets hurt."

"Well, what's in this box is for me to know and for you to find out. Just like my diary, Dr. Alchemy. You should not have taken it." Halley inched along the cave wall. Maybe she could make a run for it.

"Look, brat. I didn't want your diary. I wanted that treasure map. Now hand over the box and we can forget all of this happened." Dr. Alchemy stepped towards her, glaring.

"You forgot to say the magic word, Dr. Alchemy." She wasn't going to give him the box without a fight, but she was making him even more mad.

"Please, give me that box now." Dr. Alchemy said, slowly gritting his teeth.

Halley had no choice. She brought the box from behind her back and carefully handed it to him. She held her breath. Her stomach was tied up in knots.

"You know, I've been trying all my life to strike it rich and find gold." Dr. Alchemy said, then a shadow crossed his face. He tucked his flashlight under his arm and fiddled with the latch. "This might be the closest I get to it."

Halley tried to ease around Dr. Alchemy towards the opening of the cave.

His eyes fixated on the metal box, he didn't care about Halley any more. He reached down and flicked open the latch. "I have been waiting for this for a long time."

Halley snorted under her breath.

Dr. Alchemy lifted the lid and instantly a gush of foamy, soapy Monster Toothpaste erupted from the box and sprayed up into the cave.

"Monster Toothpaste! Monster Toothpaste!" Halley chanted under her breath and giggled at how Dr. Alchemy was reacting to the soapy foam raining down on him.

Nathan and Gracelyn emerged from the shadows giggling when Dr. Alchemy let out a high-pitched scream and ran out of the cave. They followed him and watched as he ran into the woods with his black magician cape billowing behind him.

"Think he'll come back?" Gracelyn asked.

"I don't think he knows what hit him." Nathan chuckled.

"It was only a less dangerous Monster Toothpaste." Halley shrugged. "At least he won't lose his hair over it."

"Well if he does, he deserves it for taking your

diary and trying to take the treasure of Camp Eureka." Gracelyn smirked.

"I have to admit, Halley that was the best bait and switch trick I have ever seen." Nathan turned around, picked up a different older metal box, and wiped it off. "That was a classic with a bit of a science twist."

They all looked down at the worn metal box they had unearthed a few hours earlier. They were all expecting a pirate treasure chest, but when they found this box they decided to wait to open it until they had taken care of Dr. Alchemy.

"I can't take it any longer. I want to see what's inside," Gracelyn gushed.

"Maybe it's old mineral rights papers or Confederate money." Nathan rubbed his hands together.

The three friends crouched down while Nathan pointed the flashlight at the latch.

"Here goes nothing." Halley lifted the lid, bracing herself just in case it was also booby trapped.

The box wasn't filled with money or coins or gold. It was neatly packed with a Slinky, a mood ring, a braided friendship bracelet, pet rocks, a mix cassette tape, a postcard, instructions for decoding a secret message, and an old photo of two girls smiling, arms around each other, wearing tie-dyed shirts.

Halley gingerly reached out for the picture to inspect it closer. The two girls in the photo looked an awful lot like Halley and Gracelyn.

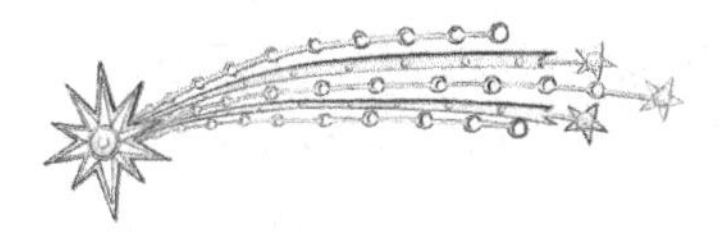

# CHAPTER 25
## 1986

HALLEY LOOKED OVER AT GRACELYN who had picked up the old postcard, turned it over, and read, "Here in 1986, we buried this time capsule after we watched Halley's Comet together at summer camp. Love, Meg and Dee."

"Wow, Halley. It's our moms in this picture!" Gracelyn looked at her with tears in her eyes. "Can you believe it? They look a lot like you and me, don't they?"

"I had no idea that Mom went to summer camp." Halley sat back and crossed her legs. "I guess back then it wasn't Camp Eureka. It must have been the young astronaut camp that Ms. Mac told me about. They must have made the treasure map I found so they could find their time capsule one day!

"Maybe this is why she named you Halley," Gracelyn whispered.

"Honestly, I always wondered why Mom named me after the comet. I never thought she even liked science."

"Maybe you are a lot more like her than you think." Gracelyn put her arm around Halley.

Halley's throat tightened, remembering how much she loved it when her mom hugged her.

"Hey, look." Nathan leaned in closer. "What do you think they're doing in the picture?"

"It sort of looks like an experiment," Gracelyn said.

Halley sat there, still unable to talk. Her mind raced with everything that had happened over the last few days.

Nathan reached in the box and unfolded a worn piece of paper. "The Friendship Experiment: If found, decode using this magic potion with your best friend."

"Oooh! Magic potion?" Gracelyn breathed. "Come on, let's try it, Halley!"

Halley stood, wiped her face, and took a deep breath. "Let's get back to camp before anyone thinks we're missing." She also didn't want another encounter with a bat or Dr. Alchemy.

They packed up the contents of the time capsule, and Halley carefully closed the latch. When they emerged from the cave, the moonlight was so bright that it lit their way back to the pathway towards Camp Eureka.

Halley noticed a few lightning bugs flickering on and off in the darkness. She had to admit, the woods seemed magical that night as they walked silently along, their thoughts flooded with possibilities of friendships and the end of summer.

"What are we going to do?" Gracelyn looked down, stopping in front of Cabin Anning. "Tomorrow is the last day of camp. Do you think we can still see each other when we get back home?"

"I'm not sure, but we'll figure it out together." Halley linked her arm in Gracelyn's and stared up at the stars flooding the dark night sky.

*Thursday*
*Dear Diary,*

*Tonight I learned that friendships can be broken and be reformed. Our friendship will never be the same; I think it will be better because friendship is the greatest treasure of all.*

*Love, Halley*

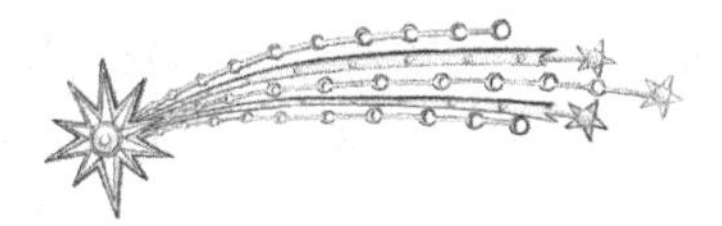

## CHAPTER 26
## THE TALENT SHOW

THE END OF CAMP TALENT show was moved inside Energy Hall due to all the rain the day before. Most of the campers sat on the floor while the parents stood along the walls talking to each other. Some parents even sat with their campers on the floor. Ms. Dee was sipping on a cup of coffee and talking to other parents. Halley hadn't spotted her mom yet in the crowd but was so happy to be sitting next to Gracelyn. She looked down at the experiment they were going to demonstrate at the show.

"I sure hope this works." Halley whispered to Gracelyn and popped her knuckles. You couldn't even hear the popping over the buzz of excitement of re-united parents and campers.

Ms. Mac hopped on stage and tapped on the microphone interrupting Halley's thoughts. It was hard to believe the start of camp was a few short days ago when Dr. Alchemy performed his Science Magic Show. Halley grinned, thinking of his expression the night before.

"Welcome, parents, to the Camp Eureka Talent Show! Your campers have been busy this past week learning all about chemistry through acid-base reactions, density, and a little magic." Ms. Mac winked in Halley and Gracelyn's direction. "The campers have been working hard on a talent show to demonstrate to you all they've learned here at Camp Eureka this summer. Enjoy learning from your campers!"

The talent show started with Lexi and Charlotte performing a skit about how to make ice cream in a bag by tossing it back and forth to each other.

Then Nathan demonstrated how to build a giant clay volcano and make it look like it was erupting.

A few other campers showed the audience how to make slime.

One of Halley's favorite performances was the teen counselors singing the entire Periodic Table of Elements *a cappella*!

Halley looked around and still couldn't see her mom. She just had to watch Halley and Gracelyn's talent, otherwise she and Gracelyn might just go home from camp and never see each other again. Halley's face felt flushed, and she popped her fingers again. Her mom was never late.

Gracelyn looked over at Halley popping her knuckles then grabbed her hand. "Don't worry, Halley." Gracelyn slowly smiled at her. "This is going to work. Your mom will get here just in time, I just know it."

"And now, our grand finale! Please put your hands together for Halley and Gracelyn here to show you their talent, The Friendship Experiment!"

Halley uncrossed her legs to stand up. They felt a little like jelly, and her feet prickled from sitting too long. Ms. Dee saw the two girls together and folded her arms, looking annoyed. Then suddenly a side door opened and sunlight flooded in. Halley's little brother, Ben, stepped in followed by Halley's mom.

Halley took a deep breath of relief because her mom was going to see their experiment. The walk to the stage seemed like slow motion as she climbed the steps towards Ms. Mac clutching the time capsule. She looked back at Gracelyn, who smiled at her and mouthed, "Don't worry." Gracelyn always had more faith in her than she had in herself.

Halley took the microphone from Ms. Mac, and it let out a loud whistle of static. The crowd collectively winced and shifted uncomfortably. Their impatience smacked her like a wave and made her even that much more nervous.

"Go Halley!" Ben waved excitedly towards her as Mom shushed him. She couldn't believe she was happy to see him. Would he go to Camp Eureka one day?

"Good afternoon," Halley said then cleared her throat. "We have learned here at Camp Eureka that science and magic seem to be very much alike." Halley looked over at Gracelyn and was glad she wasn't on stage alone.

"We are going to perform a magic trick. It will be up to you the audience to decide if it's magic or if it's science." She put down the time capsule. "But first we are going to need two parent volunteers." Halley scanned the crowd for Ms. Dee again and her mom. They were standing at the opposite ends of Energy Hall.

"Mom, would you like to come up and help us?" Gracelyn called with her hand up to her mouth.

Ms. Dee shrugged, adjusted her purse on her shoulder, and stepped over the campers to join the girls on stage.

Halley's mom looked at her watch and tapped her foot.

"And, Mom, would you like to come up and help us too?" Halley said a bit more quietly and looked towards Gracelyn for reassurance.

"I'm not going up there, Halley." Mom said loudly towards the stage. "Now get your things. We're going home."

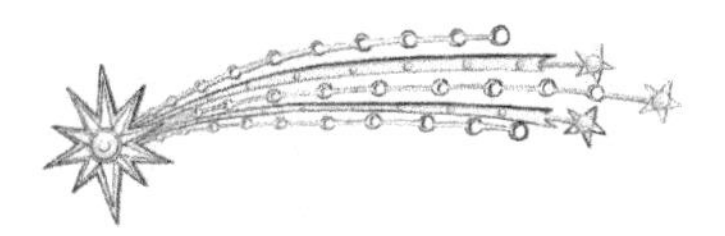

## CHAPTER 27
## THE FRIENDSHIP EXPERIMENT

HALLEY'S MOUTH DROPPED OPEN. THIS couldn't be happening. Was it because she didn't want to stand with Ms. Dee? Or did she not want to do a science experiment with Halley?

Then she felt a warm hand on her shoulder.

"Halley, let me try," Ms. Mac offered and grabbed the microphone back from Halley. "Now Margaret, I'm asking you please come forward to be a volunteer. It seems to me that the girls have discovered one of the treasures of Camp Eureka." Ms. Mac gave the microphone back to Halley and winked. Mom blushed and sheepishly walked onto the stage to stand next to Ms. Dee.

"Go on, Halley, take it away," Ms. Mac whispered and stepped back.

"Okay." Halley cleared her throat. "What I have here before you is what seems like an ordinary piece of paper. Would you agree this paper looks ordinary?" Mom and Ms. Dee nodded in agreement.

"What we have here is a magic potion." Halley motioned to Gracelyn who held up a vial of purple liquid. "This potion will reveal a secret hidden message on this ordinary paper."

Halley handed her mom a paint brush, and Gracelyn handed her mom one too.

"Could this actually be what we think it is?" Ms. Dee whispered to Mom and smiled.

"I think so." Mom snorted, sounding giddy as they both dipped their paint brushes in the potion.

Halley watched the two moms turn into kids again. She had no idea what message was going to magically appear, but she grabbed Gracelyn's hand and they both held their breath.

Mom painted the potion on the first half of the paper. Then Ms. Dee painted on the second half.

Slowly these words appeared:

SCIENCE AND FRIENDSHIP:
IT'S LIKE MAGIC, ONLY REAL.

"I can't believe it worked!" Mom laughed with tears at the corner of her eyes. "How long ago was that anyways, over thirty years ago?"

"Friends again?" Ms. Dee offered.

"Friends," Mom responded as the two hugged each other.

Then the moms looked towards their daughters and pulled Halley and Gracelyn into a big group hug, and they all jumped up and down laughing.

Halley looked at Gracelyn and breathed a sigh of relief while the parents and campers clapped at the reunion on stage.

Ben was the first to break up the hug when he ran on stage holding his own magician hat. "Halley, I missed you!"

Halley rolled her eyes and grinned at Ben's big hug that almost knocked them all over. "Hey, Ben. I missed you too."

"Maybe someday I can come with you to Camp Eureka." Ben stared at his big sister proudly. "By the way, was that really magic that revealed the message?"

"No, silly. It was science." Halley elbowed Gracelyn, and they both started laughing. Halley felt her heart lighten as they all jumped off the stage together and walked outside of Energy Hall. Mom and Ms. Dee were laughing and looking at the contents of the time capsule.

"Well, if it isn't the stars of the talent show!" Ms. Mac remarked, walking up to the girls.

"Ms. Mac, I want you to know what happened to Dr. Alchemy." Halley twisted her fingers together.

"Halley, there are always some secrets that you don't reveal, even to your best friends." Ms. Mac winked. "Camp Eureka is special in many ways, and in time, all the treasures will be revealed."

"So there really is another treasure?" Halley grinned, looking from Ms. Mac to Gracelyn.

"I guess there is always next summer to find out." Ms. Mac raised her one eyebrow, shook a box she was carrying, and then handed the box to Halley. "Here,

I wanted to give you and Gracelyn your camp take-home gifts."

Ms. Mac leaned in close and whispered, "I'll see you back here next summer, Miss Disaster." And with that, Ms. Mac turned on her heel and walked back towards Energy Hall.

Halley quickly opened the box and saw it was filled with beautiful multi-colored rocks. She closed it again and read the label that said: Camp Eureka Rocks!

Halley and Gracelyn turned towards each other and smiled.

## HALLEY HARPER: SCIENCE GIRL EXTRAORDINAIRE
## BOOK 3 PREVIEW

HALLEY SQUINTED, HOPING TO GET a glimpse of the top of the rock wall that she had started climbing. She imagined enjoying the view from the top in an attempt to forget how sore her fingers felt from gripping the gritty rock face.

The sun shone directly overhead, piercing the bright, clear-blue Texas sky.

She took a deep breath, inspected the rope securing her to the top, and found the next foothold. She set her jaw, convincing herself that today gravity was not going to win.

Why did she have the need to be at the top anyways? Maybe she got it from Dad who always wanted to climb to the top of every building, tower, or structure when they went on family vacations. She thought about Dad and how busy he had been at work lately.

She just had to keep climbing. She hugged the grainy rock and focused on the next hold. She had rock climbed before indoors but never a real rock outside. This was definitely more challenging, but Grammy always said, "Remember how to eat an elephant. One bite at a time." Grammy had a way with words, but Halley knew she would get to the top focusing on one foothold at a time.

The higher she climbed the more rock layers she passed. This rock must be sandstone that was probably formed from an ancient beach of sand and silt that was deposited in layers over millions of years ago.

It reminded her of her layered birthday cake, and how she still couldn't believe she was eleven!

"Go, Halley, Go!" Gracelyn cheered from below.

Halley smiled but didn't look down at her best friend. They had been inseparable since last summer. Now that they knew what it was like to be apart, they never wanted to do it again.

"Thanks, Gracelyn!" Halley called out and enjoyed a breeze that blew across the rock.

She always knew she could trust Gracelyn. They played the trust game as little kids. Gracelyn would close her eyes, and Halley would hold her hands and lead her around the edge of the neighborhood pool. The last time they played, Gracelyn missed her step, and they ended up falling into the pool together.

A cicada made a low, long, droning noise that sounded like a warning for how hot the day was going to be.

Strange. This part of the rock face looked different

than the layer after layer she had been seeing. What could have caused a rock a million years old to form differently than the rest of the rock around it?

"You can do it, Halley!" Gracelyn shouted from below. "There's a crack just beyond your reach that you could hold on to."

Gracelyn must have thought her change in ascent meant she was having trouble. But Halley was re-energized by the idea of exploring a new section of the rock. She had a feeling that it was different and special. What if some prehistoric creature was trapped in the rock for millions of years?

"Halley, what are you doing?"

"I just want to check something out!" Halley called out.

"Are you sure? Don't get yourself in a spot that you can't get out of," Gracelyn warned.

Halley pulled herself sideways and found a rock she could rest on for a moment. Then her gaze suddenly rested on an impression that was a little bigger than her hand. Was she just imagining it, or was she really seeing a skeleton stamped into the sandy layers of the rock?

"Gracelyn, you'll never believe this," Halley called. "I think there is a fossil up here!"

Her fingers reached out to touch it, but it was just beyond her grasp. She would have to move over a little more to inspect it. She regretted not bringing a crayon and paper to do a rubbing of the skeleton. Would anyone believe her if she said she saw a real fossil?

She adjusted her stance and reached as far out as she could to touch the fossil, but her center of gravity was off. Her fingertips were almost there, but she had to find a better foothold or she would fall.

She looked up towards the top, and it seemed farther away now that she made the detour.

An eagle swirled in the sky. Perhaps it wasn't an eagle...

It was a vulture making a long, lazy circle directly overhead. She heard the cicadas droning again, and the heat of the day made her head feel light, and her leg muscles started to twitch.

She looked down and instantly regretted it. Gracelyn seemed farther away, and the rocks below looked sharp. Falling could really hurt. Perhaps today gravity was going to win. Why did she let the fossil distract her from focusing on the top?

Little stars floated in front of her eyes, and she felt like she was going into a dark tunnel as she tried to regain focus on the fossil. She wanted to memorize what it looked like.

Halley's heart pounded so hard she could hear it in her ears. Thump. Thump. Thump.

She looked down at Gracelyn. And then everything went black.

Sign up for my newsletter at www.TracyBorgmeyer.com to find out when the future books of Halley Harper will be released!

Please consider leaving a review if you love book 2 of Halley Harper: Science Girl Extraordinaire! The rankings and reviews help with visibility to future readers. Thank you!

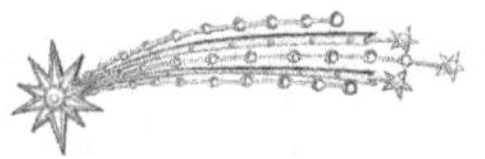

Want to get an inside look into Halley Harper's world?

Follow her on Instagram @HalleyHarper and see what science Halley is exploring now at #HalleyChecksItOut

# ACKNOWLEDGMENTS

Thank you to Mom and Dad who always give me honest feedback and constant inspiration. And to my mother-in-law and father-in-law who always remember to ask me how the book writing process is going.

To my cousins, what would I do without you? Your excitement fuels my writing passion and I love all the ideas you gave me for Halley Harper.

I truly appreciate my friend and editor Jennifer Crosswhite. She helped my spark of Halley come alive, and for that I will be forever grateful.

I am completely thankful for my talented illustrator Mel Cordan and her amazing dog, Zen. You, my friend, push me to be the best author I can be for your illustrations.

I would like to thank my family and friends who always ask me about my books and give them away as birthday and Christmas gifts. I am so honored by that gesture.

Thank you to the readers of the She Loves Science blog. Writing for you inspires me more than you will ever know.

Thank you to my kids—Allie, Andrew, and Avery—who would take turns each morning sitting on my lap while I was writing and experiment with me in the afternoons in the kitchen. I love you forever.

And most of all, I want to thank my best friend Matt, who always believes in me. Let's keep following our dreams together.

# ABOUT THE AUTHOR

**TRACY BORGMEYER** is a chemical engineer and founder of She Loves Science. She lives with her husband and three kids in Beaumont, Texas. Her life's passion is writing and inspiring parents to bring a love of scientific discovery to their kids. You can find her blogging about her science adventures at SheLovesScience.com. Her first book She Loves Science: A Mother's Guide to Nurturing the Curiosity, Confidence, and Creativity of Her Daughter was published in 2016. Her second book Halley Harper, Science Girl Extraordinaire: Summer Set in Motion was published in 2017.

She enjoys spending time with her family, writing, reading, and experimenting with recipes in the kitchen. She also enjoys cheering on the Fighting Texas Aggies.

Visit www.tracyborgmeyer.com for upcoming book releases. She'd love to hear from you at tracy@shelovesscience.com and connect on Facebook at https://www.facebook.com/shelovesscience/

## GLOSSARY

**Acid**—something that tastes sour like lemon juice

**Acid-base reaction**—a reaction that occurs when an acid is neutralized by a base

**Alchemy**—an old science thought to change ordinary substances into gold

**Atom**—a very small particle (and the name of Halley's cat)

**Base**—something with a bitter taste like baking soda

**Catalyst** – a substance that will speed up a chemical reaction

**Charles's Law**—under certain conditions when temperature increases, volume also increases

**Chemistry**—a science that explores changes to substances when they interact together

**Density**—a measure of how packed in a substance is in its containment

**Element**—a substance that cannot be broken down to a simpler substance

**Evaporation**—when a liquid turns into a vapor

**Molecules**—a group of atoms joined together

**Ozone**—a molecule made of three oxygen atoms and found in high altitudes

**Periodic table**—a table of elements organized in rows and columns

**Reaction**—when two substances react together to form a new substance

# SCIENCE EXPERIMENTS IN HALLEY HARPER BOOK 2

## SECRET MESSAGES

**What you need:**

Baking soda, water, white paper, grape juice, and two cotton swabs

**How you do it:**

1. Make the invisible ink by mixing a tablespoon of baking soda to 1/4 cup of water.
2. Dip a cotton swab in the mixture and write your message on a white sheet of paper.
3. Let the message air dry.
4. Reveal the secret message by dipping a clean cotton swab in grape juice and gently swiping it over the secret message.

**What's the science?**

The secret message appears as a result of an acid-base reaction. The grape juice is acidic and reacts with the baking soda writing to produce a color change allowing the message to magically appear! Science is like magic... but real! Visit www.SheLovesScience.com/secretmessage to watch a secret message for you!

# BEST FRIEND S'MORES

**What you need:**

A marshmallow, graham crackers, chocolate, jar, and parent supervision

**How to make it:**

1. Ask a parent to help you place a marshmallow in a jar or microwavable container.
2. Microwave on high for 30-45 seconds.
3. Carefully remove, wait to cool, and add crumbled graham crackers and bits of chocolate.
4. Stir it and enjoy!

**What's the science?**

Marshmallows are made of moisture, trapped air, and sugar. The microwave makes the trapped air hot and the air expands. The air expanding causes the marshmallow to blow up! This is Charles's Law – heating air makes it expand. See a video of this experiment at www.SheLovesScience.com/smores

CPSIA information can be obtained
at www.ICGtesting.com
Printed in the USA
LVHW021946060820
662559LV00016B/1187